# THE
# BUSINESS ANALYST'S
# SURVIVAL GUIDE

*Managing Interpersonal Dynamics
and Leveraging Knowledge of
Repeat Behavior Patterns*

## LEANN SIMONSON, CBAP

Copyright © 2012 LeAnn Simonson, CBAP
All rights reserved.
ISBN: 1466431768
ISBN 13: 9781466431768
Library of Congress Control Number: 2011918857
CreateSpace, North Charleston, SC

*With gratitude to my enduring mentors*


*Steve Ring*

*Tom McSteen*

# TABLE OF CONTENTS

About This Book..................................................................................... vii

Chapter 1
*Interpersonal Dynamics and Repeat Behavior Patterns*.................... 1

Chapter 2
*Soft-Skill Survival Tips* ........................................................................ 13

Chapter 3
*They Don't Get Business Analysis*....................................................... 21

Chapter 4
*I Am Not Getting the Respect I Deserve*............................................. 35

Chapter 5
*People Are Difficult*............................................................................. 45

Chapter 6
*Eliciting the Requirements Is Painful* ................................................ 55

Chapter 7
*Working with the Technical Staff Is Painful*...................................... 65

Chapter 8
*The Management of the Project Is Driving Me Crazy* ....................... 75

**Chapter 9**
*The Business Analysts Are Not Working as a Team* ............................ 85

**Chapter 10**
*I'm Struggling* ........................................................................................ 95

**Chapter 11**
*The Business Analysis Challenge Response Framework* ................. 107

**Chapter 12**
*Messages Not to Miss from This Book* ............................................. 117

**Appendix:**
*Individual Challenges and Possible Response Strategies* ............... 121
    *They Don't Get Business Analysis* ............................................... 122
    *I Am Not Getting the Respect I Deserve* ..................................... 133
    *People Are Difficult* ......................................................................... 147
    *Eliciting the Requirements Is Painful* ........................................... 153
    *Working with the Technical Staff Is Painful* ............................... 160
    *The Management of the Project Is Driving me Crazy* ............. 172
    *The Business Analysts Are Not Working as a Team* ................ 181
    *I Am Struggling* ............................................................................. 187

*About the Author* ............................................................................... 195

# ABOUT THIS BOOK

## WHO IS THIS BOOK WRITTEN FOR?

Whether you are a soon-to-be, new, or seasoned business analyst (BA), this book is for you. If you are a BA frustrated with "politics" and you are ready to dropkick your BA Toolkit, this book is especially written for you.

For those of you who are planning to enter the field of business analysis, the intent of this book is not to help you avoid the challenging scenarios described here. Rather, when these scenarios occur (and they will), this book will provide you with some strategies to consider as you decide how to respond.

For those of you who have only dipped your toe into business analysis, you have likely experienced some challenges, such as unclear roles or competing interests. Don't be dismayed; managing interpersonal relationships is part of the job. One intention of this book is to reframe these frustrating dynamics as expected, inherent, and solvable challenges of business analysis work. The complexity of this human interaction–heavy work is one of the reasons why the professional knowledge and skills of a BA are needed.

Veterans of business analysis have probably experienced some of the scenarios described in this book. If you have, be grateful for having real-world experiences and lessons to consider and apply in the future. Even when things go drastically wrong, you can always learn something.

Veterans of this line of work also, no doubt, have begun to see patterns. You may interpret your first challenge as the result of the characteristics of a particular work experience: "That organization was messed up." Or maybe you had a challenge with particular people: "Bob was the problem." But when it happens again in a different environment with different people, you realize that some challenges are the result of working with humans within the typical dynamics of any work environment and trying to make change happen.

Business analysis is focused on changing how business is done. Most people in a business have

> *The BA Toolkit: The knowledge, skills, techniques, and approaches used by effective BAs are often referred to as the "BA Toolkit." The BA Toolkit becomes fuller and more diverse with experience. Be sure to pull out your BA Toolkit as you manage your interpersonal relationships. These tools are multi-purpose.*

an opinion on how business should be done. It's no wonder that business analysis is especially subject to frustration. It is this realization that helps you react calmly, tolerantly, and objectively when clashes occur over how things are to be done. Don't take it personally or focus your energy on the blame game. Instead, analyze and strategize.

Interpersonal skills are critical in business analysis work. In fact, it doesn't matter if you have strong "hard" skills in conducting analysis and are armed with all the latest and greatest business analysis tools, if you lack the "soft" skills to bring others through the process. You need both hard and soft skills in this line of work. This book is focused on the softer side.

## WHAT IS THIS BOOK ABOUT?

If there is one sentence that describes what this book is intending to teach, it is this. As a business analyst, make it a habit to attentively analyze the interpersonal dynamics from your individual frame of reference, leverage the knowledge you have about repeat behavior patterns, and be deliberate in your response to any interpersonal challenges presented.

You must be effective in many areas in order to be successful in your business analysis work. But even with a strong business analysis knowledge and skill base, issues rooted in interpersonal dynamics can undermine or significantly impact your success. On the other hand, working from the power of your relationships to affect interpersonal dynamics and challenges has the potential to propel your business analysis work forward.

This book will:

- Help you customize an approach for addressing your unique challenges around interpersonal relationships. The Business Analysis Challenge Response Framework provided in chapter 11, and other strategies offered throughout this book, will provide you with aids for developing an approach for each situation; and

- Serve as a reference when interpersonal challenges, frustrations, issues, or problems arise.

## BOOK ORGANIZATION

This book is organized as follows:

- Chapter 1 discusses the management of interpersonal dynamics within your relationships and how to leverage knowledge about repeat behavior patterns to increase your success as a BA.

- Chapter 2 offers core soft-skill tips important for every BA to master.

- Chapters 3 to 10 provide a discussion of some common interpersonal challenges in business analysis and how to respond to them. I also share some personal stories from my experience.

- Chapter 11 offers the Business Analysis Challenge Response Framework to provide a structure for responding to a particular challenge.

- Chapter 12 recaps key messages from the book.

- The appendix takes the list of common challenges from chapter 1 and offers possible root causes, potential new truths, and possible response strategies to achieve the new truth.

## WHAT IS THIS BOOK *NOT* ABOUT?

Business analysis is a broad field. This book does not cover business analysis methodologies, business modeling techniques, the history of business analysis, or business analysis trends, to name a few.

For more information on business analysis, the International Institute of Business Analysis (IIBA) is a great resource. The IIBA is the de facto business analysis standard-setting and certification group and offers excellent materials, such as the IIBA's *Guide to the Business Analysis Body of Knowledge* (*BABOK*). I would encourage you to get involved with the IIBA, if you are not already.

# CHAPTER 1:
# INTERPERSONAL DYNAMICS AND REPEAT BEHAVIOR PATTERNS

Have you ever been following an analysis approach that by all accounts seemed to be the best fit for the effort at hand, yet something wasn't working? As a business analyst, you have many things to consider, analyze, plan for, and manage—not only the business problem and analysis approach but also the stakeholders, communications, relationships, personalities, skill sets, preferences, history, team dynamics, and organizational culture. At the center, you are playing an interactive role with all of these factors and participants. For any particular situation, you should ask yourself, "What are the unique human dynamics in this situation, and how can I work with these dynamics to be more successful as a business analyst?"

This book provides you with interpersonal tips and techniques from the business analysis trenches, emerging from real-world successes, observations of repeat behaviors, and lessons learned. This book also provides a framework for working through your specific interpersonal challenges.

# LEVERAGING REPEAT BEHAVIORS OR "ANTI-PATTERNS"

Some business analysis challenges tend to occur over and over again. Why is this? Consider the concept of the anti-pattern. The term "anti-pattern" was first used in reference to software patterns that are used repeatedly and thought to be effective but aren't (Koenig 1995).[1] The assertion is that by identifying these anti-patterns, better and more effective patterns can be substituted. Since 1995, many books on anti-patterns in software and other industries have been written, including anti-patterns related to analysis and requirements. Some of the common scenarios in this book illustrate anti-patterns that are pervasive in the business analysis field. Once these anti-patterns are identified, we can substitute new, more effective behavior patterns and approaches and do so quickly when we see the pattern repeat. This book will teach you how to use the powerful combination of analyzing interpersonal dynamics and leveraging knowledge gained through observation and reflection to become a more successful business analyst.

If you find yourself in a situation where you can't seem to turn the *Titanic* away from the iceberg, but you see an anti-pattern, embrace the opportunity to see it play out. If you can effect change, then by all means, this is what you want to do. But either way, it's a win. Be sure to document the lessons you learned and bring this information to the discussion the next time this anti-pattern emerges.

---

[1] Koenig, A. (1995). Patterns and Anti-patterns. Joop, 8(1), 46-48.

## MANAGE INTERPERSONAL DYNAMICS FROM YOUR INDIVIDUAL FRAME OF REFERENCE

The nature of business analysis brings interpersonal dynamics into the focus of professional concerns. You, the BA, are part of the context you must analyze and manage. You must also analyze and manage the particular people you interact with.

BAs often analyze activities within a particular scope; they determine what inputs are needed, what specific tasks are included, and what outputs or outcomes result. Below is a summary of inputs, tasks, and outputs you might consider as you manage the interpersonal dynamics in your relationships.

**Scope**

- A particular interpersonal challenge

**Inputs**

- Stakeholder analysis

- Business analysis communication plan

- Identified concerns with interpersonal dynamics

- Lessons learned from similar situations

- Known and pertinent behavior patterns

- BA Toolkit

## Activities

- Consider stakeholder analysis and the business analysis communication plan for current projects that are relevant to an interpersonal challenge

- Assess the interpersonal dynamics from your individual frame of reference

- Consider lessons learned from similar situations

- Consider known and pertinent behavior patterns and possible predetermined responses

- Analyze your current challenge, what you and others involved want to be true, and how you get there

- Adjust and establish a new set of interpersonal dynamics

- Monitor, respond to, and assess the changing interpersonal dynamics as you move forward

- Reflect on interpersonal dynamics experienced

## Outputs

- Analysis of your current challenge, what you and others involved want to be true, and how you get there

- Individual BA challenge log entries (keep a reference list), responses, and outcomes

- New reflections on behavior patterns, anti-patterns, and responses

- New reflections on lessons learned

- Updated stakeholder analysis

- Updated business analysis communication plan

- Additions to your BA Toolkit

Managing interpersonal dynamics takes conscientious attention and action, but your time and effort will pay off. Because much of the dynamics in your relationships will repeat, and you are the common denominator among all of your business analysis efforts, improvements in one case will positively impact future efforts as well.

Consider the management of interpersonal dynamics as a task to complete with every analysis effort. Interpersonal dynamics will always be there and will affect your work, and ultimately will affect whether you are achieving success for the business. In addition, you can apply knowledge of anti-patterns and how to change them to positively influence and manage the effects of interpersonal dynamics on your work.

## BUSINESS ANALYSIS CHALLENGE RESPONSE FRAMEWORK

The Business Analysis Challenge Response Framework provided in chapter 11 will help you respond to and solve interpersonal challenges you experience in your business analysis work. This framework provides a method for solving your challenges yourself, as your story is unique.

In business analysis, we clarify the business problem, need, or opportunity. We clarify the vision and the high-level business requirements. We study the as-is (current) state of the business. We specify the details of the to-be (future) state. We often need to iterate and adjust as we learn more. You can use this same process to address your interpersonal challenges.

## COMMON INDIVIDUAL CHALLENGES IN BUSINESS ANALYSIS WORK

Below is a list of challenge categories discussed in this book. Take a moment to read through the list and determine which are *your* challenges. The appendix speaks directly to this list of challenges and offers possible root causes, potential new truths to aim for, and strategies for responding to these specific challenges.

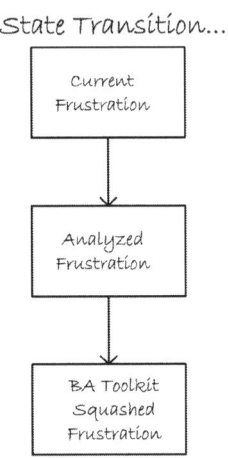

**They don't get business analysis.**

- They don't understand the importance of business analysis.

- They are hampering the business analysis process.

- They are infringing on my role.

- They are expecting me to fulfill part of their role.

- They have wrong assumptions about my work.

**I am not getting the respect I deserve.**

- They don't respect me as a BA professional.

- I am not allowed to use my professional discretion.

- They are telling me how to do my work and they don't have business analysis expertise.

- I am forced to use the same templates, approaches, and tools for all analysis, regardless of fit.

- The deliverables I am expected to create are not the right ones.

- The approach I am expected to use is not the right one.

- They see me as a note-taker.

**People are difficult.**

- The people I work with are uniquely difficult.

- People ask for my assistance while they forcibly oppose me.

- I act on what we agreed upon and then they seem surprised.

**Eliciting the requirements is painful.**

- They don't know what they want.

- They repeatedly change their mind.

- I don't have enough access to actual users or subject-matter experts.

- The stakeholders are not on the same page.

- This takes more time than anyone knows.

**Working with the technical staff is painful.**

- I find it impossible to communicate with the technical staff.

- My work is not used downstream.

- They have started working on the solution before the analysis is done.

- The technical staff are bypassing me and going directly to the business stakeholders/users/customers, and now my work is out of date.

- There isn't a professional space for me now that we have gone Agile.

- I feel like I'm leading the business stakeholders to slaughter, as the technical team is not strong.

**The management of the project is driving me crazy.**

- I am being pressured by the project manager for estimates before I've done any analysis.

- I am not given the time I need to do adequate analysis.

- The project decision makers are making bad decisions that affect my work.

- How can I do analysis when the project is sinking?

- The purpose of the project is not clear.

**The business analysts are not working as a team.**

- There is competition among the BAs.

- The BAs disagree about the correct approach.

- The roles among the BAs are not differentiated.

- Other BAs are putting me down outside of our BA team.

- There are too many "cooks in the kitchen."

**I'm struggling.**

- I am now a BA, but I don't have any training or experience as a BA.

- The work is so varied, I don't know which approach or tools to use.

- I am just burned out—this is exhausting.

You will find strategies and suggestions for responding to challenges such as these throughout this book, and the appendix responds directly to the above list of challenges.

The next chapter offers a number of business analysis soft-skill survival tips. Your success as a BA is as much determined by your soft skills as it is by your knowledge of business modeling and requirements elicitation. You must be able to communicate and relate effectively with people in order to solve business and people problems in your role as a BA.

# CHAPTER 2:
# SOFT-SKILL SURVIVAL TIPS

I used to be a software developer. I loved how the world would fade to black and I was left with the intimacy of me and my machine. The feedback was trustworthy and immediate. The feedback that comes from BA work is a little harder to decipher. The response you get could really be fueled by your subject-matter expert's broken dishwasher or unwieldy to-do list.

I used to also be a social worker. The soft skills below were important when I was a social worker, they are important now that I am a business analyst, and they are likely important in any discipline that is built on a foundation of relationships. Business analysis requires both hard and soft skills. This book is primarily focused on soft skills.

## START BY BUILDING RAPPORT.

Time is money, and we can't waste time with idle chitchat. I feel for you if this is your culture. We shouldn't underestimate the value of taking the time to build rapport. I have found that good things happen when you have built rapport with someone. People are more willing to collaborate, to offer more information and time, to give you some slack when you need it, and to recommend you. Make a mental note that it is important to build rapport with those around you. Find something in common. Show interest. Make eye contact and smile. If building rapport is not natural to you, plan to develop this skill. You will see dividends.

## VALIDATE THE FEELINGS AND IDEAS OF THE STAKEHOLDER.

Counselors will tell you that in relationships, sometimes your partner just wants to be heard, not be given solutions to the problem. Sometimes stakeholders just want to hear, "That must be difficult." Whether or not you agree with the stakeholder, train yourself to paraphrase what you hear. "So, what I hear you telling me is...." BAs routinely validate requirements. Make it a habit to validate the feelings and ideas of your stakeholders as well.

If your stakeholder offers off-topic or out-of-scope information, stop yourself from abruptly cutting the person off. I think we are trained to be the scope police in doing our analysis, but we risk making the stakeholder feel invalidated or discounted. Validate the thought and either let it float on by or place it on a parking lot list if it should be considered later. If the person belabors the point, you might need to restate the agenda or revisit the scope statement. But do so in a way that validates the person's thought, makes them feel heard, and gets you back on track.

## GO TO WHERE THE STAKEHOLDER IS.

You are a crackerjack BA and you know that great things can be achieved in your analysis effort. You have planned your approach and you set out to make the pitch. Hold up. It's important to know where the stakeholder is and start with the stakeholder's perspective in mind. Don't begin from your perspective, as it might not make sense to that stakeholder. Put yourself in the shoes of the stakeholder. Do your stakeholder analysis, connect with the stakeholder, and then make your pitch with the stakeholder's perspective in mind. In my social work days, we used to say, "Go to where the client is." I don't want to hear about single-parent support groups when I need to address the eviction notice that has been placed on my front door.

If you have a case to make, put it in your back pocket. Start by understanding your stakeholder's needs. You might develop a new understanding, adjust your case, or decide you don't know enough yet. And remember, you are there for the business stakeholders' needs, not for your own.

If you encounter conflict, stop the debate and really understand the other person. Our instinct is often to try to convince the other person of our viewpoint. Meanwhile, the other person is probably busy thinking the same thing. Really hear the other person out and then present your opinions. You'll be in a stronger position to make your assertions with an understanding of the other person. It's not about being nice. (Well, maybe it's somewhat about being nice.)

## MAKE SURE YOU REALLY UNDERSTAND THE STAKEHOLDER.

We just talked about knowing where the stakeholder is. Also, know what is important to the stakeholder. What does he or she have to gain and lose? What impact and influence does he or she have on what you are trying to do? If you have spent any time in the BA field, you know that stakeholder analysis is an important part of the work. If you are working with someone closely, it is important to go beyond a few questions in a stakeholder analysis template. We really need to know that person and what makes him or her tick. There isn't time to really know everyone in the world, in your organization, or in your customer base. But when it matters, really know the person.

What matters to this person? That we are thorough? That we keep costs down? That we don't drop tons of time into this thing? What is important to you might not be important to the other person. What individual and professional aspirations does the other person have? The better you know the other person, the more you can adjust to that person and be successful in your interactions.

If you play your cards right, you can also build rapport and understand where the other person is. That's the thing about soft skills, they build on each other. Show a lack of soft skills and the damage can ripple. If you don't understand the stakeholder, you might unknowingly interact in a way that creates a divide instead of rapport. You may focus on things miles away from where the stakeholder is.

## REMEMBER THAT A PERSON EXISTS IN HIS OR HER ENVIRONMENT.

Always remember that the other person exists in his or her environment. A person may be part of a group, family, community, country, culture, population group, and so on. A person may have any number of triumphs, struggles, wins, and losses. A person can't always close all other processing in the brain to focus on what you want to discuss.

If you are asking questions about the other person and building rapport, maybe you'll learn some details about the other person in his or her environment. But there is a good chance that there are things you won't know about. Remember this and give the other person the benefit of the doubt when potential conflicts arise. Cutting the other person slack will put you in a stronger position by not damaging your relationship. It's not about being nice. (Well, maybe, again, this is somewhat about being nice.)

## MANAGE THE STAKEHOLDER'S EXPERIENCE.

There have been times when I have been very pleased with the deliverables I have created with subject-matter experts only to realize that their experience wasn't so good. I have had the reverse happen as well. I have produced deliverables I was not happy with, yet the subject-matter experts were pleased with the experience. Good deliverables and a good stakeholder experience don't always happen together. Both take your attention. While you are managing your deliverables, be sure to also manage the stakeholder experience. Be transparent in your approach. Be responsive and sensitive to their needs and tolerances. Ask them how they are doing and for their assessment of how the work is going. Be tuned into the experience they are having and adjust as needed.

## CHOOSE YOUR PRIORITIES. CHOOSE YOUR BATTLES.

Perfectionism will drive you crazy in the business analysis profession. Let the small stuff go and focus on your priority items. This advice pertains to both your business analysis and your interactions. Does this matter? Does it matter way more to the other person than it does to you? Should you let it go because you have done due diligence? When you feel like you have encountered a battle, pause and ask yourself if this is a battle to fight or to concede. Be okay with letting some things go.

## IT TAKES A TEAM TO PRODUCE SOLUTIONS.

A solution is better when it results from many minds in synergy.

Most people are trying to fulfill their responsibilities and do a good job. Honor this and give people the benefit of the doubt. In all the organizations I've consulted in or been employed in, I can only think of a few cases where someone was purposely slacking. I remember that someone padded time into every task. She was worried about job security. I remember someone else who surfed the Internet all day and was fired. I remember another person who wandered around constantly, making the cube rounds and taking frequent breaks. But the vast majority of people I have worked with or encountered are really trying to do good work. Wouldn't you agree? When conflict emerges, assume the other person is trying to do the job in the best way he or she knows how. So are you. Now you already have something in common. Work out the interface between your two roles. Be a team player.

## BUILD, DON'T BURN, BRIDGES.

When you think you messed up with someone, try to make it right. The rippling damage might have occurred, but you can turn it around. You might need to work with this person in the future. As I've been pointing out, it's a way to improve your position in your work. And yes, it's about being nice and doing the right thing too.

If you can't make it right and the other person writes you off, give yourself some slack. Learn from the experience. Remember that you can only control what you do, how you respond, and how you make peace with the scenario. You can't control the other person's actions, responses, and thoughts. You can attempt to influence, but other people are outside your scope of control.

## DON'T GET TOO SOFT.

Soft skills are very important for a BA to possess. But be sure to maintain a balance of hard and soft skills. Sharpen your technical skills with business models. If you don't have a technical background, learn more about technology that is applicable to your environment. Even though business analysis is about technical and nontechnical solutions, technology is an integral component of business in today's world, and it behooves the BA to have some expertise in technology.

# CHAPTER 3:
# THEY DON'T GET BUSINESS ANALYSIS

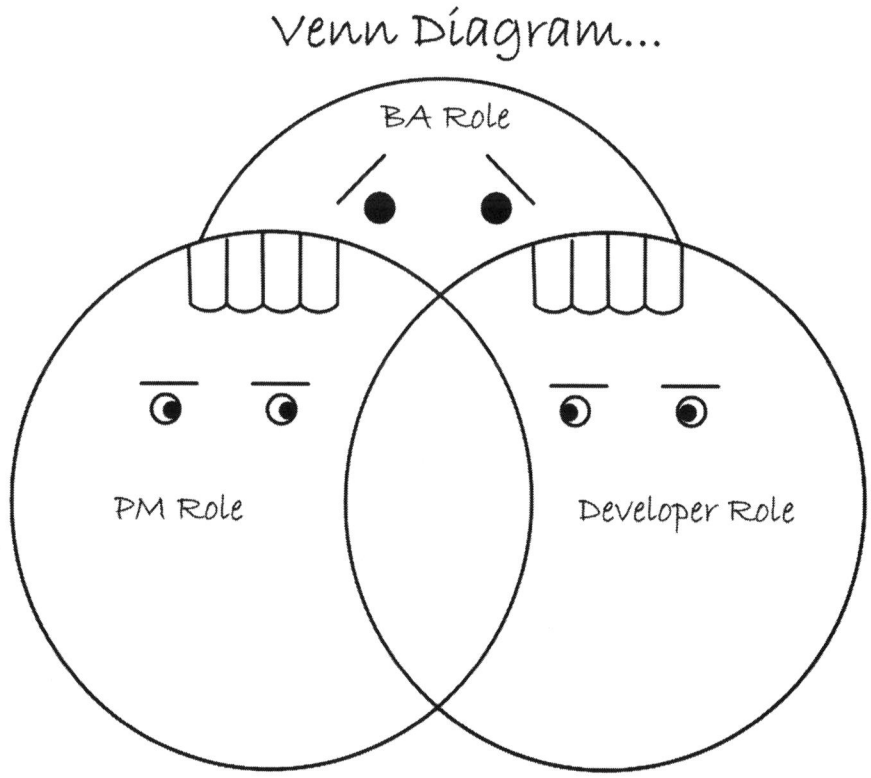

Do any of these statements ring true to you?

- They don't understand the importance of business analysis.
- They are hampering the business analysis process.
- They are infringing on my role.
- They are expecting me to fulfill part of their role.
- They have wrong assumptions about my work.

It can be frustrating when those around you don't understand what business analysis is and why it is worthwhile. As a result, we as BAs might encounter the challenge of simultaneously focusing on doing business analysis and also creating an environment in which business analysis work can be done successfully.

> *When they don't get business analysis...*
>
> - *Start by determining their drivers and motivations. Then explain how business analysis can benefit them.*
> - *If you feel like you are explaining the same thing over and over again, make it your personal challenge to improve your delivery of the message each time. The more you explain something, the better you will be at explaining it the next time. By explaining and making the case*

I'd like to offer a story from my BA experience. This story, and those offered in the chapters to follow, is structured to follow the framework you will find in chapter 11, the Business Analysis Challenge Response Framework.

## "I'VE NAILED IT."

I was once struggling with a difficult political work environment (aren't they all?) as I worked on requirements deliverables for a time-pressured project (again, aren't they all?). When the BA team and I were mid-process, the project manager (who did not have business analysis expertise) presented to the BAs his vision of where the process was headed, represented in a model. In other words, the project manager had prepared a to-be business process model. Unbeknownst to me and the other BAs, he had been working busily on analysis just like us—for the same domain, for the same project, alone in his office. When

> for business analysis, you are likely to understand business analysis in a deeper way as well.
>
> - Prepare "elevator speeches" —short speeches you can relay in the time it takes to ride the elevator. When you need to communicate on a topic, you will be ready with key points. For example, prepare an elevator speech for the questions "Why should we create an as-is business process model? Why not just jump to the to-be?"

the project manager presented his visionary model, he beamed with pride and really felt he had nailed it. The project manager then proceeded to shop his model around to all the stakeholders without enlisting the assistance of the BAs. More at issue, the business stakeholders reacted negatively to the fact that they were not involved in developing the model. I don't think this project manager got business analysis.

Before you read on, reflect on what you would have done if you were the BA in this situation.

## WHAT WAS SO FRUSTRATING ABOUT THIS? WHAT WERE THE ROOT CAUSES?

The key frustration in this scenario was that the BA team's work was usurped because there was a lack of definition around roles, responsibilities, and the business analysis process. Business analysis was not mature in the organization. Also, the BA function was controlled by a project manager with no background in business analysis.

From the project manager's perspective, he was frustrated by the lack of progress and clarity on the project. The project manager was not connected to what the BA team was actually doing and he did not make an effort to find out. Also, the BA team was not transparent and did not ensure their process was laid out for the project manager.

## SO WHAT? WHY WAS THIS A RISK OR A PROBLEM?

The *business* was not benefiting from professional business analysis.

The *project* was not benefiting from professional business analysis and the input of subject-matter experts.

The work of the BAs was dismissed and the BA team felt undermined, making them less effective in facilitating positive results for the business.

## WHAT WOULD I HAVE LIKED TO BE TRUE?

I would have liked for roles and responsibilities to have been clear. I would have liked for the BA team and project manager to have worked collaboratively together. I would have liked for the business analysis process to have been followed. The BA team wanted to be effective in their work. We also wanted to be regarded as professionals. More effective BA work results in better outcomes for the business.

From the project manager's perspective, he wanted to trust that the project was on track and achieving the right results. He also wanted to be viewed favorably in his role as a project manager.

## IF I'D KNOWN THEN WHAT I KNOW NOW…

What could I have done to improve the challenging situation described above?

I could have relayed what I believed to be sound practices and expressed what I believed to be appropriate roles and responsibilities. But hold up. Declaring the "right way" and pulling out industry references to best practices would have made a point, but it might not have been a point the project manager was ready to hear. It would have been better to first understand the project manager's motivations and drivers. It is important to develop receptiveness to your message by focusing on the needs of the person with whom you are trying to communicate. Then adjust your message accordingly.

The project manager's reputation and job were on the line. He felt like he had strong insights, which he did, and he acted on these. Because business analysis is about doing business better and determining the right solutions, business analysis needs to enter the stage along with decision makers and leaders who are doing just this, deciding how to do business better. We have to prove our worth in this arena.

The case to be made was that the other BAs and I had professional knowledge and skill that were critical to the project. My best argument wasn't "This is my turf and you should honor business analysis best practices." My best argument would have been that the project manager and the project would benefit greatly by leveraging us, the BAs. The project manager might have provided more support to the BA team and worked collaboratively with us if he'd understood the benefits this would have for him and the project.

I might have looked for an opportunity to compare the project manager's to-be model against analysis the BA team had completed. When the to-be vision obtained by following a business analysis process turned out to be different than the to-be vision created by the project manager, this would have been a teaching moment. We might have had the opportunity to compare the models and discuss why the business analysis process matters. Because the BA team believed that our approach was the right one, we could have put it to the test and explained specifically why the BA team's approach is better. If we think we know the best business analysis approach, we should be prepared to compare the strengths and weaknesses of our approach against others.

I could have also explained the risks I saw with the project manager's approach. For example:

- By bypassing as-is process analysis, the workarounds and variations of the process might go unidentified and therefore not have support via a new solution. The workarounds and variations might then continue in order to support activities not supported by the solution.

- By producing a future vision without business stakeholder involvement, the business stakeholders might feel as if they are being told what to do. They may not agree with the vision. A good relationship with the business stakeholders is critical to achieving the right results. The right results are determined by the business stakeholders.

- When roles and responsibilities are unclear and compromised, the project will suffer.

Identifying the risks of a certain approach is a way to influence decisions and ensure that decision makers are informed.

## KEY LESSONS FROM THIS STORY

- Gain an understanding of the needs and motivations of those involved, and then establish how leveraging business analysis will benefit them.

- Make the business case for business analysis. Explain why business analysis is a good idea and tie the reasons to results for the business.

- Get agreement on the approach up front.

- Be transparent throughout the process.

- Identify the risks of not doing business analysis.


## THEY DON'T GET BUSINESS ANALYSIS: AN ANTI-PATTERN

Have you seen an anti-pattern in your environment that could be attributed to people not getting business analysis? What would be a better pattern of behavior or approach?

**The "Just Build the Thing" Anti-Pattern**

I've experienced the "just build the thing" anti-pattern more times than I can count. I've observed many people taking actions based on the belief that we are wasting time with this business analysis stuff: "Let's just wing it, jump to the solution, and get it done."

It could be that the decision makers have not heard the benefits of business analysis articulated or have not seen the benefits demonstrated. Business analysis might feel like wasted time.

What would be a better pattern?

- We want others to take actions driven by the belief that business analysis will get us better solutions in the end.

- We want to ensure effectiveness by doing the appropriate business analysis for the situation. We want the right business analysis approach for the methodology being used in the environment.

What's the case for this better pattern?

- There may be a better solution, or more than one viable solution. We understand this once we understand the problem, need, or opportunity better.

- The requirements of the business may not be apparent without study. We may imbed workarounds, poor workflow, outdated processes, and other imperfections in the new solution when we skip analysis or give it short shrift. The chosen solution might fail to support the desired business capability.

If we are dealing with the "just buy the thing" anti-pattern instead of the "just build the thing" anti-pattern, we might have to work a little harder to explain why business analysis is still important. If the sponsor wants to purchase a product without doing due diligence around requirements, you might relay the following:

- We want to make sure we don't buy a fork and later determine that what we really want to do is eat soup. We want to define the "what" before determining which "how" is the right one.

- We want to make sure we are not giving up the leveraging power that comes with providing vendors with information on how the product will be used (use cases) so that we can ask them to show how the product would work to support these uses.

- The project may come under scrutiny later for how well the documentation justifies the product choice that was made.

All three of these speak from the sponsor's perspective. Reasons such as "Our process requires that we develop requirements" or "I need to better understand the choices here even if you think you already know" are not going to persuade the sponsor to think differently.

If the sponsor insists on a product purchase without developing requirements, the business analyst should document the risks relayed, as surely any fallout will and should point to absent requirements. Remember that fallout can be turned into a lesson for the next time you make the case for solid requirements before purchasing a product. Be sure to assess the results of the decision.

Take steps to educate others on why business analysis matters. The best way to make the business case for business analysis is to demonstrate success and tie this to success for the business. Ask for a small project with adequate time and resources and then make a case example out of it.

Use project failures to identify where a lack of analysis was the problem and make a case example out of failures.

Prepare your response strategy to anti-patterns such as the "just build (or buy) the thing" anti-pattern. When you see the anti-pattern start to play out, you'll be better prepared to influence the dynamics.

## A FEW FINAL THOUGHTS ABOUT NOT GETTING BUSINESS ANALYSIS

**First Impressions Matter.**

I was eliciting nonfunctional requirements (how well a solution should function) using a handy nonfunctional requirements reference list. My subject-matter experts were losing interest. At one point, someone said, "We would be in compliance with whatever IT decides." I realized that I had not set up the discussion properly. They did not understand why they should care about nonfunctional requirements. Why should these decisions be made by business people rather than technical people? I knew that the business should care about nonfunctional requirements, and I as the business analyst should have been more effective in explaining why they should care.

Consider this nonfunctional requirement. An emergency monitoring and response business needs their computer system to be available twenty-four hours a day, seven days a week. Steps should be taken to ensure that the computer system doesn't go down, even on weekends and evenings, as emergencies can happen any time. It might be fine for other computer systems to be unavailable evenings and weekends, such as the paper-pushing, during-work-hours-only company. In fact, it might be wasteful to ensure up-time and support around the clock when the computer system will not be used past the five o'clock whistle. When exploring computer system availability requirements, we as BAs might ask, "When do you need this computer system to be available to you? Is it okay if it is not available during certain hours?" If we are not on our game, we might ask, "What computer system availability requirements do you have?" This might elicit the response "We would be in compliance with whatever IT decides." The point is, there are better and worse ways to elicit nonfunctional requirements and to explain the relevance to the business. And it is important to start strong.

When I found the subject-matter experts losing interest, I adjusted my questioning to avoid technical language and to focus on benefits to the business. I had to work hard to gain their attention again and only recovered enough to complete the nonfunctional requirements, not to gain interest.

Subject-matter experts will be more patient and receptive if you set up the discussion to emphasize its value to them. Not only did I learn something about how to elicit nonfunctional requirements, I learned that the way you set up a discussion really matters. I asked myself, "What might I have said up front to have made the relevance of nonfunctional requirements clear? What might I have said to explain the importance of hitting the mark and not expending resources beyond the required level?" I might have started with a story about online sales being lost because customers were unwilling to wait for a painfully slow, overburdened computer system. Another, more impactful example is the emergency monitoring and response company that lost its ability to respond to an emergency because a computer system was down and a life was lost. I could have acknowledged that while nonfunctional requirements might seem like they are technical decisions, they impact the business, sometimes with serious consequences, and must be considered by the business stakeholders.

When you are working hard to help people understand business analysis, be sure to reflect on how an interaction went, but in addition, pay particular attention to the setup. The initial understandings, messages, expectations, and stated reasons for why someone should care can make a huge difference in how a subsequent interaction goes. Ask yourself, "What can I say up front to explain why they should care? How can I ensure that they have bought into the process before we begin? What are the right initial messages for this person or group? What examples could I use that speak to this business's domain?" Start out strong and on the right foot. The remaining journey is more likely to stay on course.

# CHAPTER 4:
# I AM NOT GETTING THE RESPECT I DESERVE

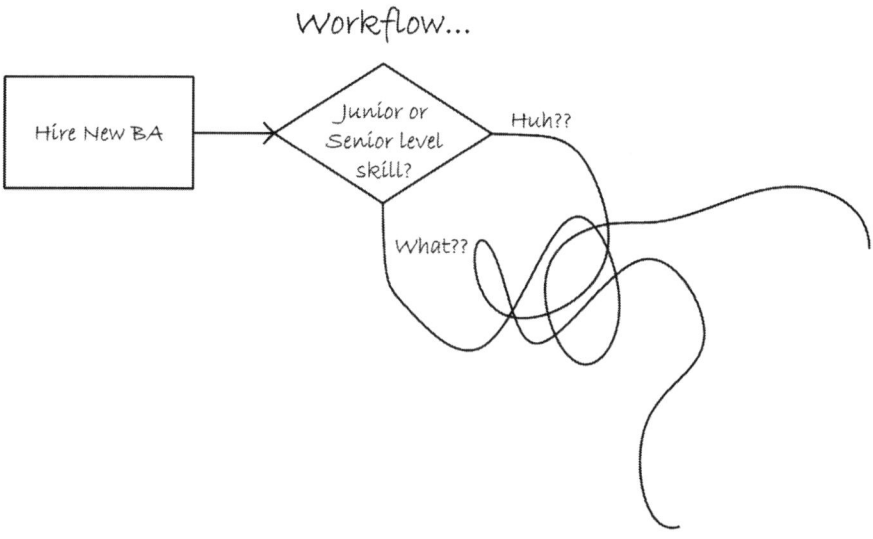

> When you're not getting respect, remember...
>
> - Respect is earned through demonstrated success. Focus on demonstrating success and the respect will follow. Strive for excellence.
>
> - Respect is easier to gain with active promotion from above (i.e., sponsors and leaders). Gain support and champions from above.
>
> - When people like you, they are more likely to respect you. Build rapport.

Do any of these statements ring true to you?

- They don't respect me as a BA professional.

- I am not allowed to use my professional discretion.

- They are telling me how to do my work and they don't have business analysis expertise.

- I am forced to use the same templates, approaches, and tools for all analysis regardless of fit.

- The deliverables I am expected to create are not the right ones.

- The approach I am expected to use is not the right one.

- They see me as a note-taker.

It can be frustrating when you feel like those around you don't respect you as a professional nor allow you to use your professional discretion. When you feel like you are not respected, it's easy to dwell on what others think: "Why don't they respect me? What are they thinking about me?" It is important to move beyond these thoughts and to develop strategies for building respect.

## "WHO'S IN CHARGE?"

I once lost control of a process modeling session I was facilitating. I was developing a process model on the front wall, eliciting information about the process from subject-matter experts. Suddenly, a developer decided to hook up a projector to his laptop in order to project a blank document on the wall. Then he suggested that the group list the process steps. Because he was viewed as more key to the effort, the group changed focus

- Respect yourself and be confident. Take pride in your work.
- Show respect to others.
- Be professional. Dress the part, act the part, and speak the part. Maintain your professionalism regardless of whether others do.
- Analyze yourself. Are your behavior and attitude such that they warrant respect? Can you make any improvements?
- Accept that respect will grow over time.

to the developer. I had to work to regain control, suggested we complete the process model, and used the information in the list to feed the model, showing that not all the facts were covered via the list. But the session was less successful than it would have been had I maintained control.

Before you read on, reflect on what you would have done if you were the BA in this situation.

## WHAT WAS SO FRUSTRATING ABOUT THIS? WHAT WERE THE ROOT CAUSES?

The key frustrations in this scenario were:

- I was not able to maintain the floor so that I could apply my knowledge and skill, in which I had confidence. The value of the process I was using was not clear to the developer. He feared we were wasting time. Also, the developer outranked me in this organizational culture. He could get away with this behavior.

- The trust the subject-matter experts initially had in my approach was lost when I was undermined by someone who outranked me.

## SO WHAT? WHY WAS THIS A RISK OR A PROBLEM?

- The takeover undermined me as a professional.

- The takeover stood to deprive the project of the value of the modeling approach, which was designed to discover more of the necessary facts than the approach taken by the developer.

- The takeover stood to deprive the business of the ultimate approach benefits, namely, business process improvements and solutions that support an improved business state.

## WHAT WOULD I HAVE LIKED TO BE TRUE?

I would have liked to have been trusted as a professional. I would have liked for there to be patience and trust in the process so that I would have felt successful in my work and so that I had the context in which I could be successful.

From the developer's perspective, he wanted the session to be successful with an approach that made more sense to him and produced results more quickly.

From the subject-matter experts' perspective, they wanted a successful session with completed documentation of their process. They were also interested in process improvements.

## IF I'D KNOWN THEN WHAT I KNOW NOW…

What could I have done to improve the challenging situation described above?

I could have declared, "Hey, I'm the facilitator!" But the likely response would have been "Not anymore!" The developer in this scenario had more clout.

What I needed to do was to gain input and agreement on the approach and the deliverables beforehand. I needed to make the case for process modeling with the developer and worked with him to define the approach and the deliverables for the session. My mistake was *assuming* he would defer to my approach simply because I was the BA and this was an analysis session.

Larger cultural issues were at play, along with issues regarding roles, responsibilities, methodologies, and so on. But the above tactic would have been more likely to achieve immediate results for the particular situation. It might also have contributed toward larger or longer-term goals.

## KEY LESSONS FROM THIS STORY

- Develop respect by demonstrating success. Why should people blindly trust you and respect you? Trust and respect are earned.

- Ask potential holdouts to join you on the path to success. Be transparent in your approach, make the case for what you are doing, and solicit agreement and input beforehand and every step of the way.

- As BAs, we are very conscious of assumptions that could affect a project. Be just as careful of assumptions you are making about the people you are working with.

# I AM NOT GETTING THE RESPECT I DESERVE: AN ANTI-PATTERN

Have you seen an anti-pattern in your environment related to a lack of respect for a BA? What would be a better pattern of behavior or approach?

## The "Template Equipped with Shoehorn" Anti-Pattern

I've experienced the "template equipped with shoehorn" anti-pattern many times in many settings. This involves an organization that requires a certain business analysis template on every project regardless of whether it fits the situation. The BA might feel as though his or her professional judgment is being undermined by inflexible rules.

Why might this be? An unbending rule that a specific template must be used might be an overreaction to a former lack of control in the types of deliverables produced. When deliverables look different each time, this can be confusing to those who receive them. Templates also ensure a certain level of rigor, and that documentation covers important information. Templates are certainly a good thing. On the flip side, however, inflexible templates might not fit the situation.

What would be a better pattern?

- A balance between using standard templates and being responsive to the analysis needs would ensure a fit for the situation. Allow templates to be customized. Or maintain different versions for different types of analysis.

- We want to ensure good work through standards, but we also want to rely on the professional discretion of the BA to determine the correct approach and deliverables.

What's the case for this better pattern?

- Every analysis effort is different. On a data-intensive project, data analysis is needed. On a process improvement project, process analysis is needed. Based on the BA's assessment, the appropriate analysis should be determined.

- A core skill of BAs is the ability to determine the right approach to an analysis effort, including the deliverables. It makes sense to leverage this expertise.

- Document the specific reasons why a template doesn't fit and the adjustments or alternatives needed in order to make the case for flexibility.

Prepare your response strategy to the "template equipped with a shoehorn" anti-pattern. When you see the anti-pattern start to play out, you'll be better equipped to influence the dynamics.

## A FEW FINAL THOUGHTS ABOUT NOT GETTING RESPECT

### Win Over One Person at a Time.

When have you felt strong respect for someone? What was it about that person that earned your respect? Some base characteristics are necessary—trustworthiness, honesty, and having others' best interests at heart. In addition, the person probably demonstrated impressive knowledge or ability, or said or did something that had an impact on you.

First, ensure that others consider you to be trustworthy, honest, and on their side. If others question your character, everything else you do will be undermined, even if you are superbly talented.

Second, develop your knowledge and skill. Look for opportunities to demonstrate them.

Third, think about individuals you are working with and what you could do to help them. Address their issue, tie what you are doing to their experience, and impact them positively.

Last, be patient. Winning over one person at a time takes time. Focus on your successes. Win respect rather than getting caught up in negative emotions. It is difficult to be your best self, develop your knowledge and abilities, and focus on others' perspectives if you're in a negative space.

# CHAPTER 5:
# PEOPLE ARE DIFFICULT

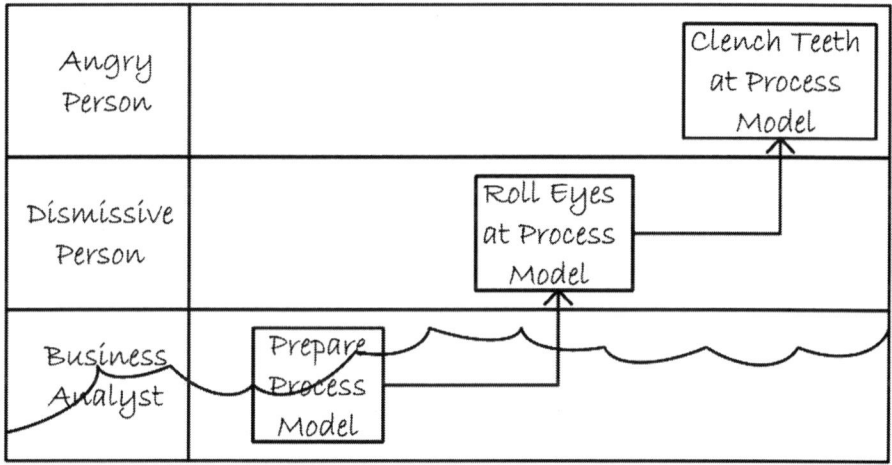

> **When people are difficult...**
>
> - Always, always be respectful. Blurting out an insult will never benefit your career, and it will be remembered.
>
> - Identify and discuss things you have in common with the difficult person.
>
> - Help the difficult person. He or she may become less difficult as a result.
>
> - If you are in the heat of a conflict, step aside and cool down.

Do these statements ring true to you?

- The people I work with are uniquely difficult.

- People ask for my assistance while they forcibly oppose me.

- I act on what we agreed upon and then they seem surprised.

It can be frustrating when you feel like you work in an environment like those shown in *The Office, Office Space, Dilbert*, or any other movie, sitcom, or cartoon that leverages the plight and absurdity of people in their natural work environments. These depictions are especially funny because they hit on kernels of truth. Each person has his or her own personality, inclinations, and temperament. Mix a bunch of people together with different characteristics from different walks of life and you will get some interesting interactions.

## "YOU SHOULDN'T QUESTION IT."

I was a new hire in a company that used a particular tool for maintaining requirements. I had minimal experience with this tool, so I asked another BA how it worked for him and what he thought of it. He responded, "You shouldn't question it. We should just use it. It is the standard."

Our relationship was off to a warm, fuzzy start. Many subsequent interactions with this person were very similar.

Before you read on, reflect on what you would have done if you were the BA in this situation.

## WHAT WAS SO FRUSTRATING ABOUT THIS? WHAT WERE THE ROOT CAUSES?

The key frustration in this scenario was that I felt unable to ask reasonable, appropriate, and necessary questions. I wanted to be able to have a discussion

> Agree to describe only the conflict and to return at a different time to resolve it.
>
> - Be sure to really listen. Don't talk over the other person or interrupt. Don't spend your interaction time thinking about what you want to say next.
>
> - Focus on the issue, not the idiosyncrasies of the other person.
>
> - If you can't make it work, consider whether you need to escalate the issue or enlist a mediator.

about our tools and methods, but instead, I quickly discovered that this type of discussion would be met with forceful resistance.

I suspect the other BA might have felt challenged or insecure around new BAs.

## SO WHAT? WHY WAS THIS A RISK OR A PROBLEM?

- Business analysis is about improving a business. We need to be free to discuss the effectiveness of how we do business analysis in order to improve.

- BAs need to work as a team.

## WHAT WOULD I HAVE LIKED TO BE TRUE?

I wanted to fit into the environment and learn the current processes and tools. I also wanted to contribute to decisions about how we did analysis. I needed this to support my ability to be effective in my work as a BA. I also wanted to further my career goals—I didn't want to settle in and be told what to do without having any input.

Thinking from the other BA's perspective, I imagine that he wanted to ensure that his seniority was respected.

## IF I'D KNOWN THEN WHAT I KNOW NOW…

What could I have done to improve the challenging situation described above?

Because of my goal to fit in and learn the process, I needed to team with this person I perceived as difficult. I also needed to not take it personally.

People are who they are. The way that people interact with you reflects who they are. Don't take a person's difficultness personally; instead, maintain your professionalism. Just because the other person is difficult doesn't mean you need to respond in kind. Control your reactions and respond in a way that moves you and the difficult person toward your mutual goals without compromising your professionalism.

Difficult people are trying to navigate life based on their own perceptions, just like everyone else. And we may all have been the difficult person from someone else's point of view. Their "difficultness" is part of how they do this, whether it is successful or not. They are not trying to be difficult in order to frustrate you; they are trying to reach their goals just as you are. The difficult person in my example was most likely maintaining his seniority and guarding against change.

Observe the difficultness of the other person, and see how the same tendencies play out in many scenarios. The difficult person described above interacted in much the same way with the other BAs as well. Notice how the difficult person uses the difficultness to navigate his or her world. This understanding will help you get into this person's head to identify motivations. If you can help the difficult person reach a goal, he or she may warm to you and become less difficult.

Respond positively if not enthusiastically to non-difficultness from a person you perceive as difficult. This is Psychology 101—reinforce behaviors you would like to see continue.

## KEY LESSONS FROM THIS STORY

- Know that you are in control of how you respond. Your response should represent how you want to present yourself as a professional. When someone is difficult, this doesn't mean you need to be.

- Know that a difficult person's difficultness is more likely to be about him or her than about you.

- Identify goals you have in common with the difficult person. Begin to work more collaboratively.

- Respond positively to non-difficultness to reinforce these behaviors.

- Know that partnering with the difficult person is more likely to help you achieve goals than confronting and challenging him or her. Honest communication and working through conflicts is important, but relationships strengthen when there are more positive interactions than negative/corrective interactions.

# PEOPLE ARE DIFFICULT: AN ANTI-PATTERN

Have you seen an anti-pattern in your environment related to the difficulty of people? What would be a better pattern of behavior or approach?

**The "Cement the Negative Trait" Anti-Pattern**

The "cement the negative trait" anti-pattern could be described as having someone pegged and pointing out to our coworkers every behavior that reinforces that judgment. The boss always talks over other people, so whenever she does it, we throw around those knowing glances. Or "Crazy Bill" never lets us down in saying something crazy and entertaining. We wait through all of Bill's useful contributions until something crazy comes out of his mouth, and then we are ready to throw around the knowing glances once again.

We are human, and humans want to make sense of their world. Detecting patterns and predictable events is a survival skill. However, we also have the capability to stereotype, pigeonhole people, and disregard their useful contributions.

What would be a better pattern?

We want to make way for positive contributions from all involved. Stereotyping people will limit our support for their full contribution to a team or project. Resist the tendency to stereotype, pigeonhole, or cement a negative trait among those you work with. Instead, watch for others' strengths, point these strengths out, and do your part to leverage the strengths of everyone you work with. Cement positive traits.

What's the case for this better pattern?

- If we build up, not tear down, those around us, we will strengthen the team. This is likely to achieve better results for the business.

- We don't want to lose out on the positive contributions someone can make because we are focused on their negative traits.

Brainstorm the strengths of your colleagues. When you observe a new strength, add it to your list. How do you see the team or project leveraging these strengths? Discuss these strengths with others, when and where appropriate.

## A FEW FINAL THOUGHTS ABOUT DIFFICULT PEOPLE

**Every Person Navigates Life Through a Different Lens.**

People behave in the way that makes sense to them, regardless of how flawed you suspect their thinking is. It can be extremely frustrating to feel that the way to do something is obvious, logical, a no-brainer…and the other person can't see this.

I must admit that I'm fairly confident in my thinking. Given different opinions, I am almost always more likely to trust my own opinion over someone else's, unless he or she has earned my respect as someone better versed in an area than me. I seem to maintain this confidence even though I occasionally find myself absolutely certain about something only to find out I was wrong. These errors in thinking help me to remember that others might be certain about something they may or may not be right about it, just like me.

When you try to persuade or convince someone, you win some, you lose some. Sometimes people don't appreciate your brilliance. Sometimes they give you credit when you know you weren't all that impressive. Sometimes people hear what they want to hear, and see what they want to see. Sometimes you break through and really connect. Sometimes you can change perceptions and understandings. Accept that all of these will happen, and that sometimes perceptions and understandings remain out of balance among people.

*C'est la vie.*

Certainly try to come to a common understanding. Certainly try to improve relationships. But when people seem difficult to you, attempt to observe and accept the other person. Then determine how you can best work with this person's personality. Resist the temptation to have that internal dialogue about how frustrated you are (which sometimes ends up as gossip around the water cooler). Instead, focus your internal dialogue on how you can build bridges, support strengths, and work collaboratively to achieve goals.

# CHAPTER 6:
# ELICITING THE REQUIREMENTS IS PAINFUL

> **When requirements elicitation is painful...**
>
> - Use what works, and swap for something else when it doesn't. Keep a well-stocked BA Toolkit. If live modeling isn't working, develop a business model off-line and validate with the SMEs (but do this only if you believe you can get close to correct).
>
> - Pay attention to the words the SMEs use and the issues they cite. Weave these words and issues into your communication. They

Do any of these statements ring true to you?

- They don't know what they want.

- They keep changing their minds.

- I don't have enough access to actual users or subject-matter experts.

- The stakeholders are not on the same page.

- This takes more time than anyone knows.

It can be frustrating when you feel like you are trying to do good work with business stakeholders who are uninterested, unavailable, unsure, waffling, or opposing you. On the flip side, as a business person, it can also be frustrating to try to get your work done when someone is absorbing a lot of your time to elicit requirements for something new and possibly a ways in the future. If you doubt the value of analysis, it can be especially hard to invest your time in it.

We often elicit requirements from people who have plenty of work already. Or maybe they are engaged, but they don't have a process for translating their needs into requirements. Or maybe the value of analysis needs to be established before conducting the analysis. For many reasons, the professional skills of a BA are needed to elicit requirements.

## "THE TIME-BOXED SESSION"

I once had the task of developing an as-is business process model with a business group, without sufficient time to do the job properly. I estimated that I would need a full day to model all of the activities of the process. But, I was unable to convince the business sponsor, who decided that I could have two hours, tops, from this busy group.

> will then know that you understand their unique requirements and are working to meet their needs.
> 
> - Ensure the SMEs know where you are going and what you are trying to create. Be transparent.
> 
> - Believe in and be enthusiastic about the importance of eliciting requirements. Your energy will be contagious. If you don't believe in the analysis, why should the SMEs?

I proceeded to get the job done. I modeled the full scope of the business process, but it was sketchy. I rushed the subject-matter experts, used lots of abbreviations, was on the edge of legibility, and ignored a lot of detail and alternative paths through the process. The result was that the work was done, but not done especially well. The experience of the subject-matter experts and the deliverable from this session were poor.

Before you read on, reflect on what you would have done if you were the BA in this situation.

## WHAT WAS SO FRUSTRATING ABOUT THIS? WHAT WERE THE ROOT CAUSES?

There were a few key frustrations in this scenario. The business sponsor did not defer to my assessment of the estimated time needed. The business sponsor's need to allot time sparingly was stronger than my rationale for the time that was needed. Another frustration was that I felt I had the knowledge and skill to conduct strong and detailed process analysis and to create a pleasant experience for the subject-matter experts, but the time constraints got in the way.

From the business sponsor's perspective, she was being asked for an unreasonable amount of time. Who meets for a full day? The business sponsor was not familiar with analysis sessions and the amount of time that is needed. Her frame of reference was other meetings that typically were an hour or two.

## SO WHAT? WHY WAS THIS A RISK OR A PROBLEM?

- I did not have adequate time to do the analysis with which I was charged.

- The business sponsor's not deferring to my judgment about effort estimates made it hard for me to be effective.

- The effort and the subject-matter experts were not able to benefit from well-done process analysis.

## WHAT WOULD I HAVE LIKED TO BE TRUE?

I would have liked for the business sponsor to trust my judgment of the time it would take to conduct the analysis in question. Also, I would have liked for the analysis to be given priority over other efforts. Ultimately, I would have liked to have had adequate time to do the analysis properly.

From the business sponsor's perspective, she wanted the session to occur, but she did not want to waste time or prioritize the analysis over other activities.

## IF I'D KNOWN THEN WHAT I KNOW NOW…

What could I have done to improve the challenging situation described above?

I could have spent more time explaining the value of conducting process analysis and the impact this work would have on the business sponsor's goals. I might have been able to move the priority of this effort higher in comparison to other activities.

If I had been unable to secure more time, I wish I would have proceeded at a reasonable pace, completed as much as I could, and maintained my standards. Then I could have gone back to the business sponsor with incomplete but strong process analysis. The subject-matter experts would have reported a good experience that was cut short, which would have helped to leverage more time. This would have also brought credibility to my estimates of the time that was really needed.

## KEY LESSONS FROM THIS STORY

- Ensure the value of the effort is clear.

- If an analysis effort does not have the priority you believe it deserves, aim to raise its priority by discussing the impacts and risks of inadequate analysis with those who set priorities.

- The business sponsor needs to be aware of the trade-offs involved in allowing insufficient time. Discuss the interplay of these trade-offs with those making the decisions about time, speed, quality, cost, completeness, etc.

- When you truly don't have the time you need to elicit requirements, consider not fully completing the analysis instead of lowering the analysis quality. Hesitate to make the analysis fit the poor time estimates, and make the case for more time. The amount of time given might be used as a baseline; you might be given inadequate time again in the future, but be asked to step up the quality and completeness. This suggestion will not always be the right option. It is important to give it careful thought. The analysis you deliver is a testament to what future analysis work will be—the impressions will live on and the constraints you were under will not be known.

## ELICITING THE REQUIREMENTS IS PAINFUL: AN ANTI-PATTERN

Have you seen an anti-pattern in your environment related to requirements elicitation? What would be a better pattern of behavior or approach?

### The "We'll Know Where We're Going if We Just Get Going" Anti-Pattern

Business analysts often refer to the functional requirements as the "what." The solution is the "how." Non-functional requirements are the "how well." But if you don't have a solid "why" as the foundation, the rest doesn't matter. Do we have a goal and are we on the same page?

Sometimes we feel like getting on with a task will illuminate the way, but instead we waste resources. It is sometimes hard to clarify a goal. We might find it easier to identify tasks we could do, even if we are not sure where we are going. We might consider our approach to be agile or iterative. We assume the goal will become clearer as we go.

But the lack of a goal or target does not make the activity agile or iterative. A goal is still important.

What would be a better pattern?

A better pattern would be to take the time to identify the shared goal because we know it is critical and worthwhile to do so.

What is the case for this pattern?

To move forward in a synchronized and focused way, we need to be clear about the destination. Otherwise, we risk wasting resources, creating conflict, and ultimately not reaching a desirable result.

Clarifying the destination or goal will:

- Give us all a target;

- Give us a measure of success;

- Save us time and effort;

- Ensure we don't build the wrong thing; and

- Ensure we don't go down a dead-end street.

Bring people together to develop a clear vision at the start of a project or effort. Vision statements, charters, and project definitions must be clear to everyone. With your group, gain agreement on the vision for each project or session.

If possible, interview each participant individually before the session to give you a pulse on perspectives, common visions, and differences of opinion. Spend time analyzing this information and use your analysis when you plan your visioning session.

If you ask, "What's our goal?" everyone on the team should have an answer. The answers should also be consistent. If not, you have work to do.

## A FEW FINAL THOUGHTS ABOUT THE PAIN OF ELICITING REQUIREMENTS

### Never Forget that Requirements Matter.

We know that requirements matter as BAs, but we need to keep this in mind when requirements elicitation gets painful. Requirements elicitation is a core part of business analysis. If we make a mistake in eliciting requirements, the ultimate solution might fail in big or small ways.

I have seen an expensive computer system fail because of one seemingly small requirement. Let me tell you about it. A solution was needed to help manage complex engineering projects across different groups. The business groups each tracked their parts of the projects separately. They relied on meetings to coordinate work across groups. A new solution would need to support the tracking of tasks and the relationship of tasks to each other across groups.

After a computer system was selected and purchased, we discovered that a task could not be tracked with multiple predecessor tasks. This one small requirement, a fact of data cardinality, was missed.

Prior to this newly adopted computer system, the users relied on disparate Microsoft Project spreadsheets to track project tasks. Microsoft Project could handle multiple predecessor tasks. After the expensive new solution was in place, projects were entered into it. Then an unforeseen reaction played out: Projects were exported to Microsoft Project for individual use and maintenance, given that it better met the needs. The result was that a computer system intended to bring uniformity and a common repository of projects backfired. The "why," i.e., the desired business outcome, was not achieved. Data was now in even more places. The situation was made worse because of a single missed requirement.

It is true that we need to guard against analysis paralysis, and that we need to adapt to settings that employ more iterative approaches. On the other hand, the above example describes how one single missed requirement can have very negative consequences. Any requirement left unspecified carries risk. Very little risk is acceptable when building software that interacts with human safety or health, but a lot of risk might be acceptable for a throwaway prototype or a feasibility study. Stakeholders must decide how much risk is acceptable given the nature of a particular effort.

# CHAPTER 7:
# WORKING WITH THE TECHNICAL STAFF IS PAINFUL

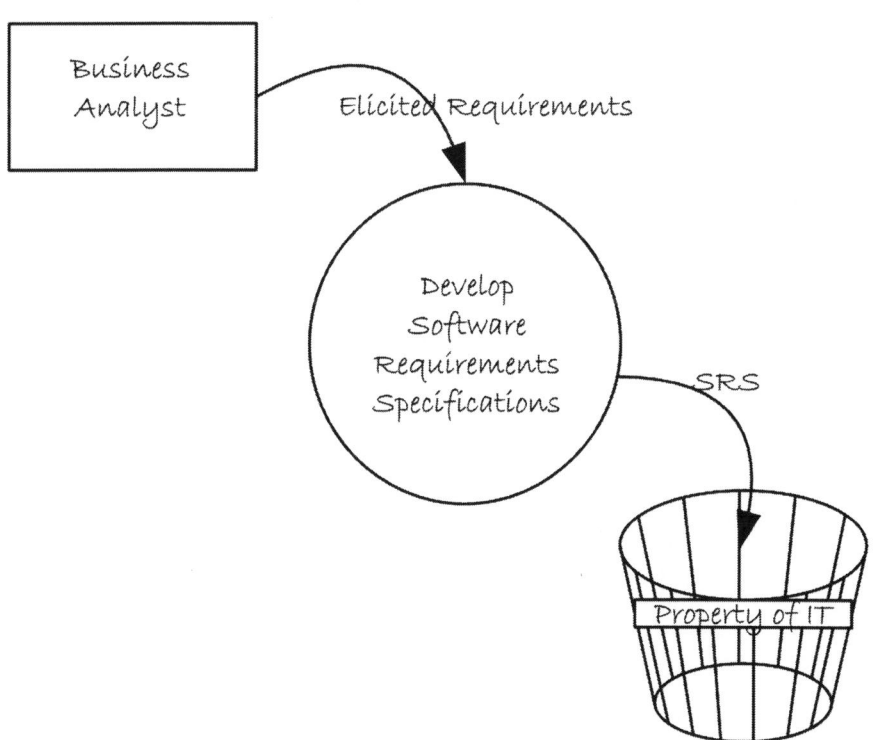

> **When working with technical staff causes you pain...**
>
> - Demonstrate what you can do that the technical staff can't do and highlight the value that you bring to the team.
>
> - Understand that technical staff are deep thinkers, working through logic and determining solutions. Standoffishness or arrogance might really be about the difficulty in shifting focus from an unsolved problem mid-stream.

Do any of these statements ring true to you?

- I find it impossible to communicate with the technical staff.

- My work is not used downstream.

- They have started working on the solution before the analysis is done.

- The technical staff are bypassing me and going directly to the business stakeholders/users/customers, and now my work is out of date.

- There isn't a professional space for me now that we have gone Agile.

- I feel like I'm leading the business stakeholders to slaughter, as the technical team is not strong.

It can be frustrating when you hit a brick wall or encounter conflict when working with the technical team.

## "BYPASSED"

I was once a BA on a project where the business liaison role was not clearly attached to the BA function. As normal, questions on the requirements came up from downstream during development. Rather than route questions through me, the developer went straight to the users. Why not cut out the person in the middle? So I was left with analysis artifacts that were inaccurate and out of date. I was bypassed.

Before you read on, reflect on what you would have done if you were the BA in this situation.

## WHAT WAS SO FRUSTRATING ABOUT THIS? WHAT WERE THE ROOT CAUSES?

It was frustrating to be bypassed. We were past the initial cut of analysis and the developer was actively working on de-

- Determine analysis deliverables with the technical staff up front to ensure fit downstream and with the methodologies used. BAs bridge the gap between business and technology, and the transitions and relationships in both directions are important.

velopment. The developer had the access to get business questions answered for his immediate questions. He no longer needed the BA to gain this clarification.

From my perspective, one big problem was that I had developed business models that were now out of date. The changes were at risk of being out of synch with the overall business architecture. Developers run the risk of being too focused on particular details and might not realize the consequences their decisions have on the larger picture. BAs are needed to ensure the big picture, the business architecture, is understood. In addition, my liaison role was compromised. The business stakeholders were unsure of their point of contact.

From the developer's perspective, to go through the BA every time would result in:

- Delays in getting answers;

- The risk of something being lost in translation;

- Unnecessary overhead; and

- A roadblock for working collaboratively directly with the business.

## SO WHAT? WHY WAS THIS A RISK OR A PROBLEM?

- The business models and other analysis deliverables were at risk of being inaccurate and less effective.

- The BA role was diminished to some extent.

- The to-be business models could no longer become as-is business models in the next release without tracking down the changes that were made downstream without BA involvement.

## WHAT WOULD I HAVE LIKED TO BE TRUE?

I wanted to be involved in all changes in order to maintain the business models and other analysis deliverables. I wanted to maintain my role as business liaison.

The developer wanted to get the current development work done correctly, quickly, and easily.

## IF I'D KNOWN THEN WHAT I KNOW NOW...

What could I have done to improve the challenging situation described above?

I could have made the benefits of updated business models that could be used in the next iteration clear to the developer. I also could have outlined the advantages of going through the BA, focusing on the benefits for the developer and the business.

Advantages to going through the BA include:

- The requirements will be up to date and ready for use by the team for the next iteration/release/solution for the same domain. The ability to update the business models for the next release or iteration would be quicker if the business models reflected an accurate baseline of the current business.

- The business models serve other purposes beyond the computer system being developed. Business models might reflect capabilities supported by a number of computer systems. Business models might include manual processes. Business models might span more than one organization. We need the full picture to make sure the computer system being developed is understood in the right context.

In environments where developers work collaboratively with businesspeople, the developer and the BA might work collaboratively and meet with the business stakeholders/users/customers together, bringing all skills to the table. I might have considered this tack when I was bypassed. I might have opted to accept that both the BA and technical staff might be part of analysis sessions or might elicit requirements through other methods. I could have relayed that when details are elicited without the BA, care should be taken to loop back so that the BA is aware, can update analysis artifacts, and can offer analysis expertise about the changes.

I should have worked to demonstrate my value as a BA both before and during development. I should have ensured my deliverables were understood and useful to the downstream developers and technical staff. I should have also gained agreement on the approach and deliverables beforehand.

You might make the case that your relationship with the business stakeholders in maintaining the requirements is put at risk when the developer steps in. But this is more about your needs. Focus on how keeping you in the loop is good for the developer. Make the case for how success as a developer must be aligned with success for the business, and that partnering with the BA is part of ensuring this alignment.

## KEY LESSONS FROM THIS STORY

- Make sure the approach and deliverables are agreed upon by the BAs and the developers. Make sure they are a good fit for the whole team's goals.

- Higher-level business architecture is important to ensure we are staying true to the big picture. Therefore an approach that works through these considerations is important even when we are developing focused pieces. Have this discussion with downstream technical staff.

- The goal is for the BA to add value to the development process and not to be viewed as a hurdle.

## WORKING WITH THE TECHNICAL STAFF IS PAINFUL: AN ANTI-PATTERN

Have you seen an anti-pattern in your environment related to interactions with technical staff? What would be a better pattern of behavior or approach?

## The "Requirements Packages Are Not Constraining" Anti-Pattern

Some technical staff might view requirements packages as ideas of what they might do in development, but not view these packages as constraining. They might assume that the analysis is now in their hands.

What would be a better pattern?

A better pattern would be to ensure that all team members know that requirements are to be aligned with business goals based on business needs. The solution should stay true to these requirements. When issues or uncertainties arise, we want to enlist the BA to update the requirements with the business stakeholders to ensure the correctness of the requirements and that the integrity of this process is maintained. The BA is often the person with the best overall view of the complete requirements.

Staying true to the requirements will:

- Ensure that the ultimate solution supports the requirements, which reflect the business vision;

- Protect traceability of design to requirements to business goal to business need;

Ensure that BAs and technical staff are on the same page regarding how requirements feed design and development. If there are reasons the technical staff are not using the requirements deliverables, determine why. Come to an agreement on how these deliverables are best determined and most useful. Discuss how requirements should be subject to a requirements change process to maintain alignment with the business goals and that they are reflected in up-to-date documentation.

## A FEW FINAL THOUGHTS ABOUT THE PAIN OF WORKING WITH TECHNICAL STAFF

### Get In the Technical Person's Head.

When I was a software developer, we did not have BAs. We, the developers, elicited requirements from the business and did the analysis. Now that I specialize in business analysis and business architecture, I am careful to think about how I would have felt if the analysis was handed to me and I was asked to design and build from it. I am also careful to consider how I would have felt if I didn't have direct contact with the business, or if I needed to work through a BA when the requirements were unclear. There are many variations of how developers and analysts work together. Having these distinct roles comes with benefits and drawbacks. BAs bring professional knowledge and skill to the work of analysis. But they also draw some of the creativity and freedom away from developers and place constraints on how they do their work.

From talking with technical staff, an often-cited analogy is the childhood game of telephone where you whisper a secret to someone, who whispers it to the next person, and so on. In the end, you learn whether the end secret matches the initial secret. Having a BA in the middle can feel like a game of telephone. Messages can get lost or changed in translation.

Here are some other complaints I've heard from developers.

- The BA said I should have seen a missed requirement in her two-hundred-page requirements package.

- The BA is stepping into design and telling me where to put a button or what colors and fonts to use, even though user interface design is not his role.

- The BA doesn't have the expertise to work through the technical details with users/customers or third-party vendors.

- The BA's process is causing me to miss my deadlines, which affects my bonus and/or raise.

Be sure to understand the technical staff's perspective. Add value to the technical staff's work, and examine how your work affects theirs. Clarify role boundaries and maintain open communications when those boundaries are unclear.

# CHAPTER 8:
# THE MANAGEMENT OF THE PROJECT IS DRIVING ME CRAZY

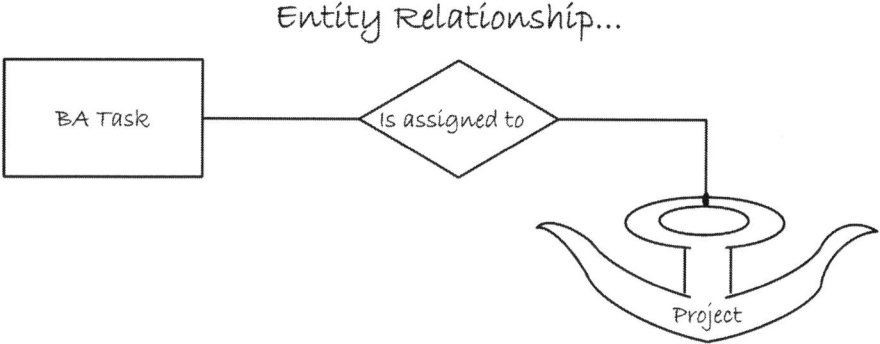

Do any of these statements ring true to you?

- I am pressured by the project manager for estimates before I've done any analysis.

- I am not given the time I need to do adequate analysis.

- The project decision makers are making bad decisions that affect my work.

- How can I do analysis when the project is sinking?

- The purpose of the project is not clear.

It can be frustrating when the project dynamics impede your ability to do analysis. Business analysis is often done in the context of a project. It is hard to be successful in the analysis effort if the project is sinking.

---

*When the project is driving you crazy...*

- *Make the case for how the project might be managed differently and how this will benefit the project decision makers. Use caution and tact so that you are not perceived as crossing boundaries.*

- *Remember that the project manager is trying to get it done while you are trying to get it right. Keep the project manager's goals and perspective in mind.*

- *Remember that managing a project*

## "CONCURRENT CONFUSION"

I once worked on a project that was in complete disarray. I was busy gathering requirements alongside a vendor who had come in to implement a recently purchased product. Even the vendor expressed surprise that the requirements had not yet been determined. The implementation was dependent on configurations that reflected the business process. My requests to correct the situation were ignored. The project manager actively continued to schedule requirements sessions and implementation work over top of each other.

Before you read on, reflect on what you would have done if you were the BA in this situation.

## WHAT WAS SO FRUSTRATING ABOUT THIS? WHAT WERE THE ROOT CAUSES?

It was frustrating when I was in the process of completing requirements, even though the

---

is not easy. Look for synergies and ways to lighten the load for the project manager, such as providing regular status updates on analysis tasks.

- Analyze how your analysis deliverables are adversely affected by project dynamics and inform project decision makers. Collaborate on how to resolve the issues.

- Speak to the project risks you see from the current project dynamics and suggest and discuss ways to mitigate or address the risks.

implementation that should be based on the requirements was underway. The project manager didn't see clear dependencies of the implementation on the analysis, despite my efforts to explain them. She believed something could be put in place and then tweaked.

The project manager was trying to get the project done, and felt like she was moving it along based on her best judgment. It was frustrating to her that the BA kept trying to stop her from moving implementation forward.

## SO WHAT? WHY WAS THIS A RISK OR A PROBLEM?

- Work done in implementation needed to be redone to reflect the to-be process still in development.

- The solution was not able to support the desired functionality because the desired functionality was unknown.

## WHAT WOULD I HAVE LIKED TO BE TRUE?

I wanted the implementation work to wait until I was done with the requirements. Furthermore, I wanted the solution purchase to wait until the requirements were done. This would have increased the likelihood that the solution and its implementation best supported the requirements.

The project manager wanted the project to be successful and to reach success as quickly as possible.

## IF I'D KNOWN THEN WHAT I KNOW NOW...

What could I have done to improve the challenging situation described above?

I could have clearly explained that the success of the project depended on implementing the best solution based on completed requirements. Shortcuts translate to more time later to correct errors.

The project dynamics seemed out of whack. I needed to make the case for why things should be different and why the dynamics impacted my ability to do analysis. But before doing this, I needed to understand the motivations and drivers behind why the project was being managed as it was. Perhaps a quick-and-dirty solution was the goal, and the risk of not meeting all the requirements was acceptable to the business. Maybe I needed to adjust my expectations for the analysis work to be more aligned with the project goals, scope, and expectations. If I needed more time and I believed it was warranted, I should have laid out my approach.

Because implementation was well underway, I could have documented the rework or ways in which the configurations didn't fit the requirements as they were completed. I could have shared these findings with project decision makers. The next project might have been managed differently as a result.

## KEY LESSONS FROM THIS STORY

- Communicate your thoughts about how the project is being managed with the project manager and/or other project decision makers. Articulate the impact of the project dynamics on your ability to do analysis.

- Ensure your analysis effort is aligned with the project goals and expectations.

- Articulate how and why analysis drives the solution, not the other way around.

- If the solution is done before the analysis is complete, be sure to point out the resulting costs. Look for specific examples of how the solution does not support the requirements in order to prepare lessons for next time.

## THE MANAGEMENT OF THE PROJECT IS DRIVING ME CRAZY: AN ANTI-PATTERN

Have you seen an anti-pattern in your environment related to the management of projects? What would be a better pattern of behavior or approach?

### The "Analysis Is a Project Plan Line Item" Anti-Pattern

Often we don't feel the need for analysis until we have identified a project. We entrap all analysis in projects, and we lack an overall map of the business architecture. Certainly it is appropriate to include analysis as work to be done within a project, but if we are unable to create the "big picture" outside of a project, we may not be building a cohesive, well-designed business architecture. If we try to accomplish the advancement of the big picture within the limits of a project, we are subject to the time pressures of the project. We might be viewed as working out of scope and wasting valuable project time. It might seem inappropriate to expend resources that cannot be tied directly to a project and its results.

What would be a better pattern?

We are better positioned in projects if we have a repository of analysis and a business architecture to refer to when a project launches.

What is the case for this better pattern?

There is a strong argument to be made for the development of a business architecture of the overall enterprise outside of projects. It is important to have a map of the why, who, what, how, where, and when of the enterprise. The why is about the drivers, imperatives and enterprise level goals of the organization. The who is about the people and organizational structure. The what is about the information and materials. The how is about the processes. The where is about the locations. The when is about the controls and events that control processes. Smaller analysis efforts can focus on pieces of the architecture to improve. This improved state can then contribute to an improved overall business architecture. Consider making the case for this enterprise level work if the opportunity does not currently exist in your organization.

Let's say that your information technology group is just establishing its technical architecture. Imagine trying to develop and implement computer systems within the constraints and expectations of a project at the same time that the technical architecture is being built and computer system components are being put into place. This certainly happens. But it is easy to see the challenge with this approach.

Conducting business analysis without a map of the big picture is similarly difficult. Without this roadmap:

- When BAs scramble to create a business architecture in the heat of projects, we risk producing inadequate, rushed analysis; and

- You may develop requirements that don't build to a well-designed business architecture. We might build a piece, and we might even build it well, that doesn't fit into the whole. Business processes might be working at cross-purposes or be redundant.

When business architecture is well established, you are then better prepared to continue to develop this business architecture via projects.

If you are in an Agile environment, this would equate to working the backlog of requirements outside of the Agile increments or sprints, as well as taking some backlog items forward within a project. Working the backlog with the business stakeholders/users/customers is important not only to prioritize requirements, but to prune requirements. You are then in a better position to identify which backlog items relate to which business function within a business architecture, which relates to which computer system function.

Think outside the project box.

# A FEW FINAL THOUGHTS ABOUT THE CRAZINESS OF PROJECTS

**You Can Get it Right and Get it Done Too, Together With the Project Manager (PM).**

A common topic in this field is the tension between the BA function and the PM function. It's no wonder because the PM is trying to get it done and the BA is trying to get it right. Both functions are critical. Sometimes the same person plays both roles and feels the internal pressure of trying to both get it done and get it right.

Even though you are likely under time pressure, take the time to not only resolve the BA/PM issues for the specific project, but to discuss the overall implications for future projects. How can you best partner to deliver the desired project results?

For instance, if the PM needs estimates as soon as possible, yet the project is a new type with no baseline analysis, collaborate on how estimates will be done for this type of scenario. Discuss the difficulty in estimating something that is unknown. Analysis, after all, begins with the discovery of the unknown. The BA might develop estimates of analysis effort with ease if the project's analysis is similar to a past project's analysis. But he or she may struggle with a new business area or a different type of project. Discuss the types of estimates, estimation methods, and types of projects. Discuss what this all means for future projects.

The best way to smooth out interpersonal tension is through communication and collaboration. Communication should occur before, after, and throughout the project among the BA, PM, other team members, and the stakeholders. If you leverage the knowledge gained from past projects during a current project and have effective and ongoing communication and collaboration, the relationship between the BA and PM is more likely to become synergistic, rather than in conflict.

# CHAPTER 9:
# THE BUSINESS ANALYSTS ARE NOT WORKING AS A TEAM

Class Responsibility Collaborator...

| Dysfunctional Team Member | |
|---|---|
| -Compete with other team member<br>-Put other team member down in front of subject matter experts<br>-Advocate for self-interests ahead of other team member's interest | Dysfunctional Team Member (Other) |

Do any of these statements ring true to you?

- There is competition among the BAs.
- The BAs disagree about the correct approach.
- The roles among the BAs are not differentiated.
- Other BAs put me down outside of our BA team.
- There are too many cooks in the kitchen.

It can be frustrating when there is dissention among the business analysis team members. This situation makes it especially hard to do your work because the team dynamics are working against you. Not only is this bad for your team, the dissention is probably visible to those outside your team as well. Achieving business goals is undoubtedly put at risk.

> When the BA team isn't working well together...
>
> - Research team-building strategies and try some out.
>
> - Analyze each team member's motivations. Find a way to establish common goals and shared benefits.
>
> - Having varied opinions, thinking, personalities, knowledge, and skills is a strength of a team. Leverage the varied contributions of the team members.

## "ECLIPSED"

I remember feeling eclipsed by another BA. I was on a project, but I was not the lead BA.

During a session break, a subject-matter expert approached me to discuss some of the requirements. I began to interact, but then the lead BA sprinted over to take control of the discussion. She positioned her body to face the subject-matter expert and put her back toward me. I was rendered irrelevant.

Before you read on, reflect on what you would have done if you were the BA in this situation.

## WHAT WAS SO FRUSTRATING ABOUT THIS? WHAT WERE THE ROOT CAUSES?

This scenario frustrated me because I was immediately viewed as the lesser analyst, even though I was the lead BA on other projects.

---

- Turn conflict into creative tension. Say "We see this differently, let's explore why we each feel this way."

- Bring in a neutral third party to help facilitate the resolution of your team conflict. Connect with your HR department for resources.

- If you have ever felt eclipsed or thwarted by a fellow BA, reflect on and log these feelings. Carry this with you for when you are in a position to impact another BA.

From the lead BA's perspective, she did not want messages relayed that might be inconsistent with where she was taking the analysis. She also did not want to be out of the loop.

## SO WHAT? WHY WAS THIS A RISK OR A PROBLEM?

I felt undermined in my ability to act as an analysis expert with this subject-matter expert. This had the potential to create an impression of my having lesser capabilities and to affect future interactions with this person.

## WHAT WOULD I HAVE LIKED TO BE TRUE?

I wanted to be allowed to interact with the subject-matter expert as an expert analyst. I wanted to maintain my effectiveness as an analyst and to have the respect of both the subject-matter expert and the other analyst.

The other analyst was likely feeling ownership over her analysis effort and wanted to make sure it remained in her control and was on course. She also wanted the subject-matter experts to see her as the expert analyst for the project.

## IF I'D KNOWN THEN WHAT I KNOW NOW…

I could have found a tactful way to ask the other BA not to take over unless I said something that needed correcting. In addition, I could have found a tactful way to relay that support for the contributions of other analysts is important because:

- Not doing so can lead to resentment, embarrassment, and erosion of the team; and

- The business stakeholders could lose faith in specific analysts, the effectiveness of the team, or the business analysis process.

If the lead BA felt as if she might lose control of the analysis direction or not be apprised of the discussion, the expectation could be set that non-lead BAs who are approached by subject-matter experts be empowered to act as experts, but that the lead be consulted, enlisted, and updated as appropriate.

## KEY LESSONS FROM THIS STORY

- When team conflict emerges, determine the root cause of the conflict and the motivators of each team member.

- Discuss points of contention among BA team members, aim to resolve them, and discuss a method for resolution in future similar scenarios.

- Strive to show respect among team members inside and outside the team.

- Lift up the whole team as expert analysts in the eyes of the business stakeholders where possible.

- Work through team issues when not working with business stakeholders and, when possible, before working with business stakeholders.

## THE BUSINESS ANALYSTS ARE NOT WORKING AS A TEAM: AN ANTI-PATTERN

Have you seen an anti-pattern in your environment related to BA teams? What would be a better pattern of behavior or approach?

**The "Business Analysis Façade Team" Anti-Pattern**

I was once on a team of BAs that followed the work of a "solutioning team." The solutioning team gathered high-level requirements from the business. In this case, it meant gathering a prioritized wish list of functionality and other requirements all over the board. I think of this team as a "façade" that hid the real BA team. The real BAs were then expected to detail these requirements and determine how the requirements fit together. The problem was that the time with the business had already been burned, which I call the "burn factor." The business stakeholders complained that they were being asked the same questions again once the BAs were engaged. The two teams were not acting as one, yet they shared the same space and overlapped significantly.

It is likely that organizational design decisions were not based on an understanding of the true dynamics and that the BA team was not seen as the customer liaison in the organization.

The result was conflict between the two teams. There was also conflict within the BA team over how to detail a fuzzy needs list not organized by functionality without bothering the business stakeholders.

What would be a better pattern?

We want to ensure that we have clear roles and responsibilities that are not redundant, and that business stakeholders experience a smooth, non-redundant process.

What is the case for this better pattern?

Establishing clear roles and responsibilities among the various groups that conduct some form of analysis or requirements elicitation is vital to make sure handoffs are effective, subsequent work is based on prior work, and the business stakeholders' time is not wasted.

## A FEW FINAL THOUGHTS ABOUT BUSINESS ANALYSIS TEAMS

**Move your current As-Is Team to a future To-Be Team.**

There are certain differences I have observed between business analysis teams that seem to have "gelled" and those that are rife with conflict and competition. Some of the characteristics I have observed in teams that have gelled are:

- Roles are clear among BAs (I am a business analyst, you are the business systems analyst, and we know the difference in our environment);

- The analysis is partitioned among the BAs and specializations appear (I'm the process modeler and you are the data modeler);

- Analysts consult with each other on approaches and conduct deliverable reviews;

- Analysts support each other as a team (I'll capture details while you develop the business model. I'll review your deliverables and you can review mine);

- The business analysis team is successful and is viewed as such by the business stakeholders; and

- The BAs are provided with training and supported in their involvement in industry and trade activities.

Some of the characteristics I have observed in teams rife with conflict and competition are:

- Unclear roles and jockeying for the lead position;

- A wide variety of approaches and the tendency to step in and take over because of a perceived "better way";

- Analysts who won't help each other;

- Analysts who tear each other down rather than build each other up;

- The business analysis team is viewed unfavorably by those outside the team;

- Training is not provided to the BAs;

- A lack of involvement in the business analysis industry and trade activities; and

- A lack of regular meetings among BAs.

Move your current As-Is Team to a future To-Be Team. Strengthen your team and strengthen your career. A stronger team will propel you forward as a BA.

# CHAPTER 10:
# I'M STRUGGLING

*Functional Decomposition...*

```
        ┌──────────────────┐
        │ Conduct Business │
        │     Analysis     │
        └──────────────────┘
                 │
     ┌───────────┼───────────┐
┌─────────────┐ ┌─────────────┐ ┌─────────────┐
│Plan Analysis│ │   Elicit    │ │   Conduct   │
│  Approach   │ │Requirements │ │ Stakeholder │
│             │ │             │ │  Analysis   │
└─────────────┘ └─────────────┘ └─────────────┘
```

> **When you are struggling...**
>
> - Know that you are your own best advocate.
>
> - Remember that business analysis is not easy.
>
> - Know that everyone always has areas of growth to pursue. No one ever reaches perfection.
>
> - Take charge of yourself as a BA. If you are not the BA you want to be, transform yourself. It's never too late to become the BA you want to be.

Do any of these statements ring true to you?

- I am now a BA, but I don't have any training or experience as a BA.

- The work is so varied, I don't know which approach or tools to use.

- I am just burned out—this is exhausting.

It can be frustrating when you feel like work is a struggle and the outlook is grim. Maybe you have lost some faith in yourself. Maybe you've lost faith in the field of business analysis.

Pretend you are a neutral party analyzing your own situation. Or ask a fellow BA to help you analyze your job challenges. A shift in perspective or looking at a situation through more than one perspective can help. In research, there is a concept called "triangulation." When multiple studies are conducted of the same thing, comparing their findings and considering the similarities and differences can help to refine the results. Considering multiple perspectives on your individual challenges can bring clarity. It can help you chart a course to a better place.

## "THE MATRIX"

In business analysis, we often compare one set of data to another set of data using a table or a matrix. We might compare roles along one axis against functions in another axis to specify permissions in the cells. The RACI matrix is a popular one that you can easily Google if you haven't encountered it yet. Or you can build a matrix of one ill-defined set of something against another equally ill-defined set of something. This is what I did. Live. Big session.

Now, I might feign inexperience, but this wasn't the case. This was not very long ago. Or I might

- Know that struggles really do make you stronger in this line of work.

- Expand and deepen your skills by seeking out a balance of both new and repeat types of work.

- Conduct a 360-degree review about yourself on your own. Ask those you trust all around you, such as your manager and coworkers, for their assessment of your strengths and weaknesses. What are their suggestions for your growth?

blame lack of planning. Also not true. In fact, I planned for hours with the project team. We might have fallen victim to "groupthink," where a group is too in synch and leads itself down the wrong path, fully in agreement. But I was the consultant. I was the so-called expert analyst. I must admit that I just made a bad call.

Luckily, I had had previous successes with this client, and thankfully, I was successful afterward. But during this session, we set aside the process modeling we had been doing due to time constraints. We decided to compare a list of functional areas we found in a database against activities they performed. Then we were going to determine common activities across these functions. Confused? So were they. Partway through, I wanted to bail and move abruptly to process modeling. The project manager wanted to save the exercise and was far more active than she should have had to be. It failed. It was a waste of time. We continued in this futile analysis attempt for a couple hours.

Before you read on, reflect on what you would have done if you were the BA in this situation.

## WHAT WAS SO FRUSTRATING ABOUT THIS? WHAT WERE THE ROOT CAUSES?

My frustration with this experience was with myself. I believed at the point that it happened that I was a bit of an expert. I did not want to believe I could fail in such a visible way. But I flat-out failed.

In retrospect, I can see some issues that I could not see beforehand. The concepts we were comparing did not come organically from the subject-matter experts. We pulled these concepts from a database.

We compromised our approach because of very limited time. We used an untested strategy. We should have had a contingency plan, and then we should have used it. Hindsight is twenty-twenty, as they say. Usually, when I consider lessons learned, I think, "Yeah, should have seen this coming." But in this case, I really went in feeling prepared. I've been "matrix-shy" ever since.

## SO WHAT? WHY WAS THIS A RISK OR A PROBLEM?

- We did not get the results we wanted.

- I was embarrassed and I questioned my abilities. This compromised my ability to be effective.

- I risked fallout with a client.

## WHAT WOULD I HAVE LIKED TO BE TRUE?

I could focus on technique strengthening or contingency planning when considering the above scenario, but I would like to focus instead on being more resilient when things go wrong.

I predict this will happen to me again. Probably not with a matrix—I am unlikely to make this particular mistake again—but with some other new, grand idea. It may even be with a new business person who hasn't seen me succeed in other ways.

What specifically do I mean by being resilient when things go wrong? I mean that I don't immediately question my ability to continue doing analysis. I can say, "Hey, this failed as analysis occasionally does. This is why I think it did, and these are the lessons I can carry forward."

You might develop a "when things go wrong" response plan. Developing one without an actual failure in mind might help you be more objective.

For example:

1. Blame the other analyst? No....Quickly get a new client if you are a consultant? No, not that either....Analyze the lessons learned. Do this with others if possible.

2. Relay and discuss the lessons learned.

3. Admit you messed up.

4. If you can determine you messed up mid-stream, switch to contingency plan.

5. So, you should have a contingency plan.

6. Recall all of your successes and feel good about your body of work.

7. Carry an expectation that every so often you will fail. When you do, tell yourself, "There it is."

## KEY LESSONS FROM THIS STORY

- Don't let one failure bring down your confidence.

- Consider lessons learned from failures in the context of all your successes.

- Remember that the best insights come from failures.

- Have contingency plans and change course when something isn't working.

- Don't compare one ill-defined set of concepts against another ill-defined set of concepts live with a large group. And if you find yourself doing this, don't continue for two hours.

**Where Are You?**

Even the most seasoned BA can identify growth areas. Identify what you do and don't do well. What challenges or frustrations do you have with yourself? Why? The interpersonal relationship you are analyzing is the relationship you have with yourself.

**What Do You Want to Be True?**

Paint a picture of where you would like to be as a BA if this challenge or frustration about yourself were not true. Why would you want this? What specifically do you mean by this?

## What Is the Path to Get There?

Chart a course for how to get there. Can you get training? Can you work on a project that will give you this experience? Can you work under another BA? Can you practice this work at home or as a volunteer? Can you connect with other BAs via discussion groups or communities? Can you develop thicker skin? Can you take an incremental approach and improve in one small area? Can you accept your faults and strengthen what you do well?

Business analysis is about moving from the present to the future. Where is your future as a BA?

Pay attention to the parts of your job you enjoy and feel good about. Some would name facilitation; some avoid facilitation at all costs. Some are big-picture; some are detailed. Some lean toward technical work; some lean toward the business side. Business analysis is a big space. Be conscious of the work that brings you joy and consider ways to do more of this specific type of work.

In some cases, you might need to move on. Try not to quit too soon. Make an effort to turn things around. But try not to quit too late, after damage is done. You might need to cut your losses and move on to a better fit. If you decide it's time to leave, secure the next opportunity before you leave your current job. Knowing that you are working on a new plan can help you endure what currently doesn't fit.

# I'M STRUGGLING: AN ANTI-PATTERN

Have you seen an anti-pattern in your environment related to a personal struggle? What would be a better pattern of behavior or approach?

**The "Don't Make Any Sudden Moves" Anti-Pattern**

"I'm just going to fly under the radar until I can decide what to do. I won't ruffle any feathers. I won't make any sudden moves. I hope things will get better, but I can always bail."

Why would I do this? Self-preservation. Let the other guy take the heat. Or maybe I don't know what to do.

When I do this, I am not working to make my environment better. I am also not functioning as a professional. This is not a good career strategy. I may not need to ruffle feathers, but I should not tolerate a bad situation or not make any effort to change it.

What would be a better pattern?

The better behavior would be to actively assess the situation and work to make improvements.

What is the case for this pattern?

Doing so will:

- Make my *current* work better and more effective;

- Make my *future* work better and more effective;

- Allow me to mentor others;

- Allow me to feel empowered by having a positive impact on my environment; and

- Acknowledge that BA work is challenging wherever I go. Addressing my struggles with my current work will make me a stronger BA.

# A FEW FINAL THOUGHTS ABOUT INDIVIDUAL STRUGGLE

**Live a Reflective Life.**

If you are struggling but can commit to working on improving your situation, reflection is a powerful method. Keep a notebook by your bed, carry one with you, and be mindful about observing your world. Then, take the time to review your reflections, which might include challenges, observations, or other thoughts. If you are puzzled by something, then research the issue and talk to others. By being reflective, you are bound to make improvements in your situation the next time around as you will be armed with new strategies, thinking, and/or techniques.

If you feel like you need to find a new path, being reflective will help you work through this. Maybe there is an alternative way to define your job. Maybe there is another opportunity with your current employer. Maybe you need to move on to other employment. Maybe you need to go back to school. Maybe it's not so bad in retrospect.

Reflection will help you work through your challenge. Writing it down will help your mind from spinning out of control about the challenge. Go back and read what you wrote a month ago or a year ago. Has your situation improved, deteriorated, or stayed the same? What does this tell you? Certainly seek out help and insight from others, but also seek out help and insight from yourself through reflection.

# CHAPTER 11:
# THE BUSINESS ANALYSIS CHALLENGE RESPONSE FRAMEWORK

## SOMETIMES IT HELPS TO HAVE A FRAMEWORK.

In my former career in social work, I was once part of a discussion among social workers who were debating the validity of dream analysis. Dream analysis involves determining meaning and direction from one's dreams. One social worker was an ardent proponent of the technique. He used dream analysis with clients in his clinical social work practice. Some of the other social workers were skeptical that meaning could be derived from dreams. Yet another social worker said, "I use dream analysis occasionally and it doesn't really matter if there is validity to it or not. Dreams serve as an archetype for working through a client's problems." An archetype, framework, or model can be useful in organizing thoughts and actions. The more frameworks you create and use, the more easily their form occurs to you.

In this spirit, I offer you a simple framework as a conceptual aid to help you respond to interpersonal challenges in your business analysis work. The Business Analysis Challenge Response Framework described below is a logic model to define the problem, identify the outcomes you want, and determine the steps that are necessary to achieve the outcomes. In business analysis, we work with models and frameworks in order to work through business complexity. Similarly, we can work through the complexity of a challenge, frustration, issue, or problem with the framework below.

As stated in chapter 1, the Business Analysis Challenge Response Framework should feel familiar to you as a BA. It is a framework for moving from a current as-is challenging state to a future to-be improved state of being. The essence of the framework is to use your BA knowledge and skill to respond to and reduce or eliminate challenges to make way for more effective business analysis work.

The three primary questions of this framework are

1. What is my current challenge?

2. What do I want to be true?

3. How do I get there?

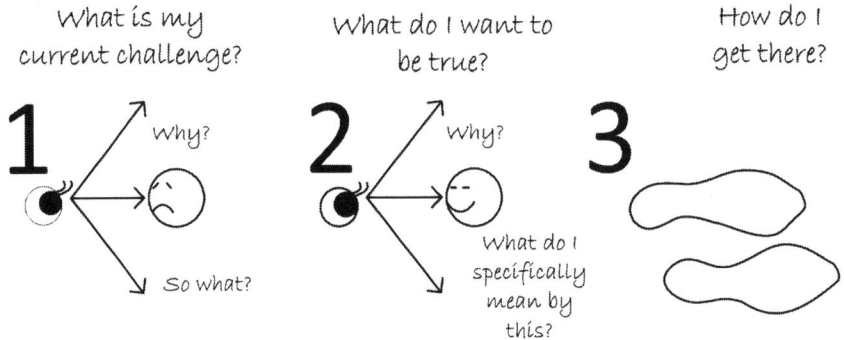

We will step through these primary questions below. The stories throughout this book have followed this same pattern. Other questions and considerations are provided below under each of these three primary questions.

## THE BUSINESS ANALYSIS CHALLENGE RESPONSE FRAMEWORK: QUESTIONS

### Primary Question 1: Ask Yourself, "What is My Current Challenge?"

1a. Name your challenge, frustration, issue, or problem and commit to addressing it.

Begin by forming a statement that crystalizes your challenge. This might be easy, or it might take some work if you haven't thought through the source of your challenge. If you are having trouble forming a statement, write it in whatever clunky rough form it occurs to you and then refine it.

Once you have a clear statement, determine if you can commit to addressing the challenge to improve your situation. If you feel like you cannot make this commitment, log that your response is to do nothing. Periodically return to the information you have logged to determine if you have new motivation to respond, or if new conditions should be evaluated.

If you feel that you can commit, work through the process below.

1b. Ask "Why?" Then, ask "Why?" again.

Once you name your challenge, ask "Why?" to examine the root cause of your challenge. "Why" is a magic word. Why does this challenge exist? Why am I in this current situation? What are the drivers or reasons for this challenge? What are the reasons from my vantage point, and what are the reasons from the vantage point of others involved in this situation? When you come up with a statement as to why, consider this new statement and ask "Why?" again, as long as it is fruitful to do so.

You might be thinking of root cause analysis you have done, perhaps with a fishbone diagram. Exactly. Apply your business analysis knowledge and skill to your challenge. You might end up with a new core understanding of your challenge. Be a persistent two-year-old and ask, "Why? Why? Why?"

1c. Look at your challenge and ask, "So what? Why is this a risk or a problem?"

Probe and dig until you have a full picture of the challenge and what will happen if it remains unaddressed. Ask the question "So what?" Ask similar probing questions, such as:

- Why do I care?

- What are the impacts of this challenge on me and on others involved in this situation?

- What are the risks and long-term implications of not addressing the challenge?

- What are the impacts on my business analysis work?

- What are the impacts to the business?

Once you explain a thought, ask yourself again, "So what?" Keep asking as long as you have more to discover. In business analysis, we decompose processes. Here, we are decomposing a challenge, frustration, issue, or problem.

You might end up deciding that the challenge doesn't have an impact you care about. Or the challenge might resolve itself with time, and it might not be worth the effort to address it. Or you might discover concerning impacts you hadn't yet thought about. You may find new resolve to address the challenge. Be a challenging teenager and ask, "So what?"

**Primary Question 2: Ask Yourself, "What Do I Want To Be True?"**

2a. Refine and name what you want to be true.

Unless you enjoy complaining about your challenge and want to keep it around, think of how you can replace it with a better situation. Consider questions such as:

- What do I want to be true? Describe this improved situation.

- How would others involved in the situation feel about this new truth?

- Do I need to adjust the vision to represent a common goal among those involved?

- Knowing the root causes of my challenge, am I addressing the true source of my challenge?

- Would this benefit me and others, or am I after something where I win and others lose?

- How would this benefit the business?

This might be easy or it might take some thought. As in question 1, if you are having trouble forming a statement, write it in whatever clunky rough form it occurs to you and then refine it. Imagine it so you have a destination. Be a dreamer.

2b. And then ask "Why?" yet again.

Ask why you want what you want. Is there something more meaningful behind your initial thought of what you want to be true? By asking "Why?" you might establish a new understanding of what you want to be true. You might ask additional questions such as

- What is the reason for wanting this new truth? What is the benefit?

- Am I sure it's worth it?

- Am I clear on my true goal?

- Do I need to refine my goal?

- Why would others want my vision to be true? What's in it for them?

You might be thinking of an analysis technique called "the five whys." Exactly. Use your BA Toolkit as you dream and create a vision for the future.

2c. Look at your vision and ask, "What specifically do I mean by this?"

Be specific and use your analysis of the challenge to help you paint a picture of the future. Keep asking, "What do I mean by this?" or, "What specifically do I mean by this?" as long as you have more to describe. Consider how others involved in the situation would answer these questions. Break down your vision into actionable and achievable goals, both for you and for those you hope to engage. It will be easier to figure out how to get there if you have a clear, detailed picture of where you are going. Dream in color, dream in details.

**Primary Question 3: Ask Yourself, "How do I Get There?"**

3a. Chart your course to a better situation.

Once you have a picture of where you are and would like to be, the next step is to determine how to get there. Think through the possibilities and prepare contingency plans. You might build a happy path, as we say in business analysis. You might also build some alternative paths as you anticipate how others might respond and as the situation unfolds.

3b. Determine if you need to secure new resources.

Knowing where you want to go or what you want to make true, consider whether you have the resources to make it happen. If not, you may need to secure these resources as part of your efforts. Ask yourself:

- Do I have the needed time?

- Do I have the needed knowledge?

- Do I have the needed skill?

- Do I have the needed support?

- Can I leverage the knowledge and skill of others?

3c. Then, make it so. Take action.

If you are overwhelmed, identify some next steps and see if you can make headway on something small. Know that you are the best person to manage the attack on your own challenges.

3d. Then, iterate.

There is a good chance you'll need to iterate through your analysis to refine your understanding. You might return to your analysis after interacting with those involved, or after taking the few first steps. Were you basing your understanding on assumptions that weren't true? Adjust your plan. Adjust your understanding. Rework or refactor your challenge response.

## KEEP A BA CHALLENGE LOG

Make a habit of jotting down individual challenges as they occur, even if the response is not readily identifiable. Later, reflect on these challenges and work through the Business Analysis Challenge Response Framework. Simply writing it down and knowing that you will think about it later can be empowering and may help you let go of it for the moment.

Update your log as you respond to a challenge and adjust your plan of attack. Make observations and reflect on your situation. This log will not only help you in your current situation, you will also be creating a reference for similar challenges in the future. A log such as this is a powerful tool.

## GETTING UNSTUCK

We often get stuck in a problem or challenge, or in our complaints. We also get stuck when we identify what we wish were true but we fail to develop a path for getting where we want to be. And maybe the biggest sticking point is that we don't take action. The basic steps described above provide a framework for moving from a frustrating or challenging scenario to a better situation. These steps might even serve to get us "unstuck."

# CHAPTER 12:
# MESSAGES NOT TO MISS FROM THIS BOOK

Business analysis is a discipline focused on business improvements and the identification of the best solutions for business needs. Business analysis work can be very complex and involved and is made more so by the human element. The way you manage the human element has the potential to reduce or increase your effectiveness independent of the quality of the analysis skill, knowledge, tools, and techniques you use to conduct analysis.

This book has stressed some important messages related to the human element. These messages are revisited below.

## USE YOUR BA TOOLKIT TO CRUSH YOUR INTERPERSONAL CHALLENGES.

You have ready knowledge and skill to leverage as you manage the interpersonal dynamics within your relationships. Be clear about where you are with these dynamics, where you want to be, and how you are going to get there. It is then important to act. This is what we do all the time as BAs—we determine how to move from an as-is state to a to-be state.

All the training and projects in the world won't automatically make you a strong BA. The strong BA pays close attention to the results and nuanced dynamics of a situation in order to build strength as a BA. The strong BA also very conscientiously uses and expands his or her BA Toolkit.

## MAKE THE CASE.

You might have noticed that throughout this book you are urged to make the case. If you want things to be different and you think you know the "right" way, provide the facts that will get people on board. Tie these facts to results for the business. If you don't have the facts to back up your ideas, or if you can't tie what you are proposing to results for the business, you might need to rethink your position.

## BUT HOLD UP! UNDERSTAND YOUR STAKEHOLDER'S NEEDS AND GOALS BEFORE YOU MAKE THE CASE.

People will not get on board if the case you make isn't in step with their needs and goals, or if it doesn't make sense from their perspectives. This requires that you understand their perspectives, goals, and needs first so that you can customize the message. By doing this, you might even change your thinking.

## VALUE LESSONS LEARNED.

If things don't go well, there are always lessons learned. I'm not advocating that you send a project off course in order to learn more lessons. Lessons will arise quite naturally. But take advantage of the lessons you can extract from every experience—especially those efforts that go off the rails.

## VALUE ANTI-PATTERN DISCOVERIES.

Some bad choices happen over and over again. We keep thinking something is a good idea and it really isn't. When you can't stop the train from veering off course but you see a pattern repeating that you think you have seen before, be sure to observe the pattern and its results. Develop a business analysis anti-pattern replacement that you can add to your BA Toolkit and polish up an elevator speech around it.

APPENDIX
# INDIVIDUAL CHALLENGES AND POSSIBLE RESPONSE STRATEGIES

This appendix lists the common individual challenges in business analysis work found in chapter 1 and suggests possible: 1) root causes, 2) new truths, and 3) response strategies to help achieve these new truths. These elements loosely follow the Business Analysis Challenge Response Framework presented in chapter 11. The full framework is intended for use with real situations in which your analysis will build on other analysis in a meaningful way. The elements chosen for this appendix are intended to trigger ideas for your use.

# THEY DON'T GET BUSINESS ANALYSIS: CHALLENGES AND POSSIBLE RESPONSE STRATEGIES

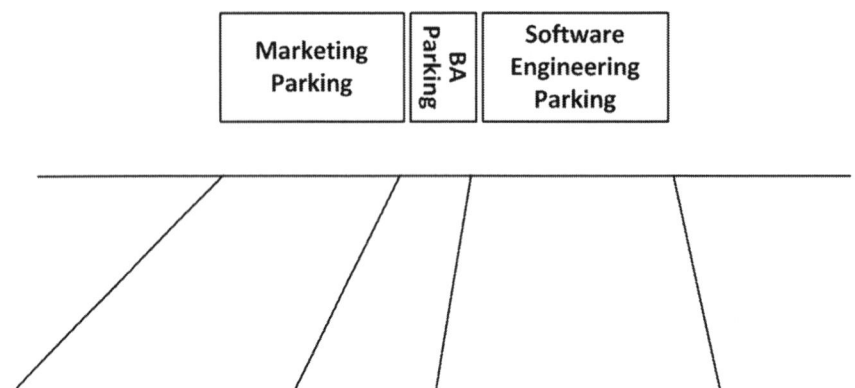

**They don't understand the importance of business analysis.**

1) A Possible Root Cause

    - Business analysis is new to the organization.

2) New Truth

    - Business analysis is on its way to becoming a mature function in the organization.

3) Possible Response Strategies

    - Provide industry information about the business analysis function.

    - Make the business case for how enlisting a skilled BA will help business efforts and people succeed.

- Study and understand the organization's culture. Make a business case for why, where, and how business analysis should fit in, what has to change, and what will be the same.

- Ask to be part of the development of a Business Analysis Center of Excellence (BACoE) in your organization.

- Analyze all of the interfaces of the business analysis function with other functions, such as project management, marketing, software development, software testing, and so on. Work with others to define roles, responsibilities, relationships, handoffs, synergies, etc. at these interface points.

- Adjust your expectations to accept the learning curve and growing pains of the development of business analysis in the organization.

4) Another Possible Root Cause

- The effectiveness of business analysis has not been demonstrated.

5) New Truth

- The effectiveness of business analysis has been demonstrated.

6) Possible Response Strategies

- Communicate your successes.

- Increase your knowledge and skill in order to be more effective.

- If there is resistance to the techniques used, ask to demonstrate your skill on a small project that is not time-critical or high profile.

- Highlight when organizational or project *success* can be traced back to solid business analysis work. Highlight when organizational or project *failures* can be traced to a lack of solid business analysis work.

- Stay conscious of the goal to demonstrate business analysis effectiveness with every analysis effort.

7) Another Possible Root Cause

- Analysis has been conducted by various roles in the past and it is assumed that a lot of people can take care of the analysis.

8) New Truth

- The business analysis role with expected competencies is an established role.

9) Possible Response Strategies

- Provide industry information about the business analysis role and expected competencies.

- Demonstrate professionalism and competency, and own the role.

- Make the business case for why a specialized business analysis role will bring a needed level of knowledge and skill to the organization.

- Consider and address role conflicts. Determine how to compensate for drawbacks of a specialized business analysis role. For instance, you might need to increase communication and collaboration to address the effects of handoffs.

## They are hampering the business analysis process.

1) A Possible Root Cause

    - Others think they have a better way and don't believe in the business analysis process.

2) New Truth

    - Others can see the value of the business analysis process over other approaches.

3) Possible Response Strategies

    - Be prepared to explain the merits of your business analysis process and practice your delivery beforehand. Be sure to stay flexible as you obtain feedback and address concerns.

- Ask others to evaluate your business analysis process for improvements. Consider elements of their approach that could be incorporated. Someone who is involved in the development of a process is more likely to accept it.

- Collaborate and gain agreement on the business analysis process that will be used for a specific effort prior to using it. It will be easier to have this discussion prior to the launch of the effort than when you are actively using the process.

- Track and tell stories in which a project *succeeded* because of the use of a business analysis process. Track and tell stories in which a project *failed* because of the lack of a business analysis process.

- Look for role conflicts under the surface or negative impacts of the business analysis process on others, such as others' losing a responsibility, creativity, or control. If identified, work to resolve these conflicts.

4) Another Possible Root Cause

- They are unaware that there is a business analysis process.

5) New Truth

- They understand and support the business analysis process.

6) Possible Response Strategies

- Conduct a walkthrough of your business analysis process and why it exists.

- If others accept the value of your business analysis process, ask for their support and promotion of the process.

- Be transparent as you are using your business analysis process. Describe what you are doing and why with each step. Make visible the full picture of how you will get to the desired end deliverables or results. You might develop and share a process model of your business analysis process.

**They are infringing on my role.**

1) A Possible Root Cause

- They don't see the value of my role.

2) New Truth

- They understand the value of the BA role.

3) Possible Response Strategies

- Discuss roles and responsibilities directly and collaborate with others on their definitions. Also specify the value of each role, including the BA role, to the business.

- Focus on how you can demonstrate the value of the BA role with each interfacing role, and work hard to achieve this. You will demonstrate value in different ways as you work with the business stakeholders, the developer, the project manager, and so on.

- Identify something that can readily make the business case for the BA role. Select something easy to accomplish quickly that clearly impacts the business, i.e., *low-hanging fruit*, and work to achieve this.

- Identify a gap that can be filled by the BA role.

- Adjust your expectations regarding how quickly your role will be embraced as you build respect over time.

4) Another Possible Root Cause

- They are unaware that they are infringing on my role.

5) New Truth

- Everyone understands each role and agrees on the role differentiations and responsibilities.

6) Possible Response Strategies

- If awareness is the issue, an open discussion about roles and responsibilities might be the remedy, or at least a good start.

- Discuss industry standards and examples of pertinent roles and the differentiations among roles. Be sure to consider the culture, methodologies, and philosophies in place in your environment.

- Shift their thinking from "I need to work with the BA" to "I want to enlist the BA because it will benefit me to do so" by proving unique value.

7) Another Possible Root Cause

- They think the business analysis activities are their turf.

8) New Truth

- All turf disputes are resolved. Everyone agrees on role differentiations and responsibilities.

9) Possible Response Strategies

- Discuss disputed activities directly and relate to roles and responsibilities.

- Express why it is important to resolve turf issues, such as activity duplication and the negative impact on the project or organization.

- Each organization defines (or fails to define) roles and responsibilities according to its culture, resources, methodologies, philosophies, etc. Be conscious of the need to be flexible, and be ready to adjust your own definitions of who should do what.

- Think through the impact of the dispute on the other roles. Is someone losing a job, control, or an activity they are good at? Consider how to respond to these impacts.

- Escalate the issue to the appropriate decision makers, if necessary.

**They are expecting me to fulfill part of their role.**

1) A Possible Root Cause

   - There is a lack of resources to fill the other roles outside the business analysis scope, such as quality-control testing.

2) New Truth

   - Sufficient resources are secured to fulfill roles outside the business analysis scope.

3) Possible Response Strategies

   - Express why securing more resources to fill these roles will benefit the decision maker and the organization. Express how these benefits will outweigh the costs.

   - Research best practices and the knowledge and skill used by the missing role. Express the knowledge and skill you don't currently have related to this missing role.

- Explain that your business analysis tasks suffer when you are spread too thin with tasks outside your scope. Explain why this affects the quality of the business analysis results.

- Explain how your focused business analysis role without the extra responsibilities will benefit the organization.

- Consider whether you might want to shift your thinking and accept a blended role. If you do, evaluate the professional development opportunities that this brings and request the training that you need.

4) Another Possible Root Cause

- There is a lack of definition around roles and responsibilities.

5) New Truth

- There is clear definition around roles and responsibilities.

6) Possible Response Strategies

- Develop a list of roles and their responsibilities as a group. Work through misunderstandings and different views to come to one common set of definitions.

- Model the work in question with a process model, identifying roles and responsibilities per process or activity. Develop an as-is business model and a to-be business model, with the subject-matter experts (SMEs) being those people representing all roles in focus.

- Identify what you do agree on first, and set aside areas of disagreement. With common ground and more context, come back to the areas of disagreement.

**They have wrong assumptions about my work.**

1) A Possible Root Cause

    - There is a lack of information about the business analysis function and wrong assumptions have been made.

2) New Truth

    - Information on the business analysis function is widely available in the organization. Wrong assumptions about the business analysis function are quickly addressed.

3) Possible Response Strategies

    - Ask others about their assumptions about business analysis. Contemplate each specific assumption, why it exists, and how to correct a wrong assumption with words and/or actions. Document this information for later reference.

    - Be transparent about your business analysis approach and deliverables every time.

    - Increase communication and collaboration with others. Suggest more regular team meetings if needed. Connect informally more often to provide updates.

# I AM NOT GETTING THE RESPECT I DESERVE: CHALLENGES AND POSSIBLE RESPONSE STRATEGIES

**They don't respect me as a BA professional.**

1) A Possible Root Cause

   - They don't know what I do.

2) New Truth

   - They know what I do as a BA and appreciate the value I bring.

3) Possible Response Strategies

   - Educate others on what a BA does. Enlist a non-BA with whom you work to attend a BA training or presentation with you.

- Introduce key decision makers to the IIBA and its literature and guidelines.

- Be transparent in what you do. Lay out your approach and the deliverables you are producing.

- Keep the goal of demonstrating value for the person you are interacting with. Keep the team, the business stakeholders, and the business at the forefront of your thoughts. You might be producing impressive deliverables, but don't lose sight of where those around you are at with your process.

4) Another Possible Root Cause

- There is some history with BAs that has not been positive.

5) New Truth

- The baggage from the past no longer holds us back. A new relationship has been formed between the BAs and those with whom the BAs work.

6) Possible Response Strategies

- Prove that the BA team, by way of new members and/or newly developed knowledge and skill, is not the same as before. Name what was then and what is now.

- Hold a meeting to gather concerns and input about the BA function and relay plans for the future.

- Build a new history of success and accept that this takes time.

7) Another Possible Root Cause

    - There has been some history with me that has not been positive.

8) New Truth

    - I have transformed my relationship with others in my work.

9) Possible Response Strategies

    - Identify what specifically happened and have a conversation with those who were negatively affected. Sometimes this is enough. Be honest and open about the history and explain what you are doing to turn things around.

    - Consider any positive shifts in relationships, even small shifts, as building success. Keep the momentum going.

    - Consider these issues as your individual growth challenge for your current job and your career.

10) Another Possible Root Cause

    - They don't understand the knowledge and skill that are necessary to do business analysis.

11) New Truth

- They are aware of the knowledge and skill it takes to do business analysis.

12) Possible Response Strategies

- Teach some of the techniques and ask others to try them out in a mock session. Those simple modeling sessions are not so simple once you are holding the marker.

- Be transparent and explain your approach and technique. Explain why you have chosen to use them.

- Actively add to your BA Toolkit. Work to continually improve your knowledge and skills in order to demonstrate more knowledge and skills.

**I am not allowed to use my professional discretion.**

1) A Possible Root Cause

- The business analysis role and responsibilities are ill-defined in my organization.

2) New Truth

- My business analysis role is well defined and it comes with an expectation that I will apply my professional discretion.

3) Possible Response Strategies

- Begin with a discussion of roles and responsibilities and work with others to define them. Specifically discuss each role's freedom to act.

- Plan and prepare for everything--every interview, every session, and every walkthrough. Conscientious planning increases the likelihood of success.

- Deliver your messages with confidence.

- Develop and show advanced expertise and skill. Others will begin to look to you for your professional expertise and skills.

- Analyze the impact of using your professional discretion. Does this take away from another role? Address the impacts you discover.

4) Another Possible Root Cause

- The business analysis capability has not been developed in my organization.

5) New Truth

- The business analysis capability is mature in my organization.

6) Possible Response Strategies

- Research and share how other organizations are building a Business Analysis Center of Excellence (BACoE).

- Research BA capabilities, starting with the IIBA Competency Model. Share your findings with others, and build a plan to develop the capabilities as a team.

- Be involved in the greater BA community and seek out conversations with other BAs.

- Stay on top of trends in business analysis.

**They are telling me how to do my work and they don't have business analysis expertise.**

1) A Possible Root Cause

- They don't understand that business analysis is a profession with a body of knowledge.

2) New Truth

- They defer to me and other BAs for our expertise.

3) Possible Response Strategies

- Make your expertise real for someone by positively affecting his or her world as a result of good BA work.

- Actively build skills and learn new techniques. Explain the approaches and techniques you are using. Demonstrate knowledge and skill.

- Introduce others to the IIBA *BABOK*.

- Explain a range of options that could be used. Explain which option you are inclined to use and why.

- Get certified as a Certified Business Analysis Professional (CBAP).

**I am forced to use the same templates, approaches, and tools for all analysis regardless of fit.**

1) A Possible Root Cause

   - Others are unaware of the need to adjust the approach to the situation.

2) New Truth

   - Others are aware of the need to adjust an approach depending on the analysis situation and its unique dynamics.

3) Possible Response Strategies

   - Affirm the value of standard templates, approaches, and tools. Use them where they fit, and explain where they need to be adjusted and why.

- Educate others on why approaches need to be adjusted. Create some "what if" scenarios to illustrate why adjustments are needed. For example: What if you find that subject-matter experts are explaining their process according to what the process protocol is *supposed* to be rather than what it is because of management in the room? You might then use individual interviews and observation to augment the approach.

- Discuss the core competency of being able to determine the right approach, bringing in the IIBA's explanation of this competency.

4) Another Possible Root Cause

    - Approaches have been all over the map and standards have become very important in the organization.

5) New Truth

    - There is a balance in our use of standards with the need for customizing the approach.

6) Possible Response Strategies

    - Determine whether some templates and standards should apply in every case for some components of an analysis effort.

    - Suggest the use of templates and standards as *guidelines* in some cases.

- Suggest the allowance of variances or alternatives. If needed, suggest a governance process around these variances or alternatives if this level of control fits the organization's culture.

- Create a feedback loop regarding how the templates and standards worked and didn't work on real assignments. Work to improve them.

**The deliverables I am expected to create are not the right ones.**

1) A Possible Root Cause

    - Those who determined the deliverables don't understand the ramifications of shoehorning all analysis into certain types of deliverables.

2) New Truth

    - It is understood that the BA needs to match the right deliverable to the situation, while maintaining a level of consistency.

3) Possible Response Strategies

    - Provide a summary of the benefits and drawbacks of the deliverables chosen and other deliverable options, such as text vs. diagrammatic models. Educate others as to why certain deliverables in certain situations are a mismatch. For example, a mismatch would be a data-centric project that will result in a new database which is limited to requirement statements or process models (and does not include a data model).

- Determine if you have the leeway to create the deliverables you believe are most appropriate along with the required deliverables, and prove their usefulness.

- Consider whether different deliverables are needed for different audiences.

- Determine how to balance consistent deliverables with the needed flexibility to adapt deliverables to the nature of the analysis.

4) Another Possible Root Cause

- They don't understand the value and importance of business models.

5) New Truth

- They appreciate the value and importance of business models.

6) Possible Response Strategies

- Explain the specific facts gathered by each type of model, and why using textual statements won't capture these facts and their relationships as well.

- Consider producing both business models and the textual equivalent as alternative views and/or for different audiences.

- Create business models as your working documentation to produce well-formed requirement statements, user stories, use cases, questions, or whatever is required as a deliverable. Share your secret as to how the business models helped you produce and structure the required deliverables.

**The approach I am expected to use is not the right one.**

1) A Possible Root Cause

   - Those who determined the approach are not aware that they are choosing the wrong approach.

2) New Truth

   - I am allowed to determine the appropriate approach based on my expertise and the situation. There is a belief that I know how to determine the right approach.

3) Possible Response Strategies

   - Engage those who are constraining the approach in a conversation about alternative approaches. Detail and share benefits and drawbacks of the approaches being considered. Be sure to gain an understanding of why they are constraining the approach.

- Consider whether required approaches can be augmented with additional work to get at the right analysis or questions.

- Bring in the IIBA's view that a BA should have the core competency of being able to determine the right approach.

4) Another Possible Root Cause

- The approaches to date have been too varied and there is a lockdown on approaches.

5) New Truth

- Approaches follow a common methodology, but with room to adapt to the situation.

6) Possible Response Strategies

- Identify the stable components of a methodology and the components that need to be flexible. Discuss this information with key decision makers.

- Delineate how the restricted approach can miss key analysis, fail to answer the right questions, and ultimately miss the mark in terms of business results.

- Work to put a method in place for improving and communicating about the business analysis methodology and approaches used. For instance, hold regular methodology meetings, set up a discussion board, document and share approach strengths and weaknesses on live projects, etc.

**They see me as a note-taker.**

1) A Possible Root Cause

- The BA role is not viewed as a professional role.

2) New Truth

- The business analysis role is considered to be a professional role.

3) Possible Response Strategies

- When required to take notes, analyze the notes after the meeting. Then turn around observations and analysis findings, and maybe a straw business model.

- Identify risks or gaps in the approach being taken, and suggest business analysis services you can offer to eliminate or reduce these risks or fill these gaps.

- Suggest that you facilitate an upcoming requirements session and describe the approach you will take and why.

- Produce professional, polished results wherever you can, even if you feel they are "notes."

4) Another Possible Root Cause

- My role conflicts with the role of others and I've been relegated to note-taker.

5) New Truth

- I am serving in a professional BA role with professional responsibilities, and all role conflicts have been resolved.

6) Possible Response Strategies

- Begin by working with others to iron out role conflicts in order to make room for your role as a BA.

- Consider whether there is more than one BA function, perhaps in marketing, Information Technology (IT), or on the business side. Discuss whether any duplication exists and how these roles are distinguished.

- Do a gap analysis of the team's work and determine whether you could meet any gaps.

# PEOPLE ARE DIFFICULT: CHALLENGES AND POSSIBLE RESPONSE STRATEGIES

**The people I work with are uniquely difficult.**

1) A Possible Root Cause

    - We don't understand where the other is coming from.

2) New Truth

- We understand each other.

3) Possible Response Strategies

- Choose a non-work topic to discuss and begin to build rapport.

- Ask questions about the other person's viewpoint before providing your own.

- Give reasons, background, and explanations to raise understanding.

- Name each of your goals and discuss how you can work collaboratively to meet both of your needs.

- Attend training on how to work more effectively with others.

4) Another Possible Root Cause

- There is a personality conflict.

5) New Truth

- We accept each other's differences.

6) Possible Response Strategies

- Find a lighthearted, perhaps self-effacing way to acknowledge the conflict. Work together to develop a way that you can embrace or even leverage each other's differences.

- Make a list of assumptions you have about the other person, and then challenge and test them. Maybe your assumptions are wrong.

- Take on a difficult personality as a challenge to sharpen your BA skills. How can you work effectively with this person? What approaches will work with this person?

7) Another Possible Root Cause

- I've been told I am difficult to work with.

8) New Truth

- Others say good things about working with me.

9) Possible Response Strategies

- Listen to feedback, especially repeat feedback, and determine how you can grow as a person.

- Give people permission to give you feedback by expressing interest in constructive criticism.

- Take time to imagine the other person's perspective, needs, and goals. Ask yourself how you can assist the person in meeting these goals.

- Offer apologies when appropriate.

- When eliciting constructive criticism, remind yourself of your strengths and that everyone has weaknesses. Take pride in the fact that you are rooting out weaknesses and improving yourself.

**People ask for my assistance while they forcibly oppose me.**

1) A Possible Root Cause

    - Others are balancing a desire to work with me with an overloaded job, and as a result they shortchange the analysis process.

2) New Truth

    - Others work with me to make room for adequate analysis time and to adjust to an agreed-upon approach.

3) Possible Response Strategies

    - Relay that a decision to shortcut requirements elicitation raises the risk that requirements will be incomplete or have errors and the right solutions may not be chosen.

Provide project examples to support this. Discuss how much risk due to incomplete requirements is acceptable to the stakeholders for the specific situation.

- Relay that time saved now by giving requirements short shrift is likely to be more expensive later when ill-fitting solutions need to be corrected or refactored.

- Be sensitive to the subject-matter expert's constraints. Be flexible and creative in your approach. Would another time, place, or method work better given the constraints?

**I act on what we agreed upon and then they seem surprised.**

1) A Possible Root Cause

    - People forget what we agreed upon.

2) New Truth

    - People retain an understanding of what we agreed upon. They also have a reference for these decisions.

3) Possible Response Strategies

    - Document understandings and distribute to all affected.

    - Call out when a decision is different than what was decided previously.

- Summarize decisions and ask for confirmation of the decisions at the end of the discussion.

- Regularly review plans and other guiding documents with those who are making decisions or carrying them out.

# ELICITING THE REQUIREMENTS IS PAINFUL: CHALLENGES AND POSSIBLE RESPONSE STRATEGIES

**They don't know what they want.**

1) A Possible Root Cause

    - The business stakeholders are unclear about their goals.

2) New Truth

    - I help the business stakeholders clarify their goals.

3) Possible Response Strategies

    - Ask "why" when the business stakeholders express a need as a strategy to get beyond solutions and to get to the true need. And then ask "why" again.

- Work to build consensus among decision makers. Bring them together as a group.

- Ask what the business stakeholders specifically mean by the goal expressed. What are the implications?

- Research similar businesses for strategies and then share findings.

4) Another Possible Root Cause

- They are clear about the goal, but don't know how to get there.

5) New Truth

- I help the business stakeholders determine a path for reaching their goals.

6) Possible Response Strategies

- Follow a business analysis process. Identify where the business is and where they want to be. Help them identify how to get there. For example, develop as-is and to-be business process models. Then conduct a gap analysis between them to identify what is needed to get to the to-be business process, i.e., the solution requirements.

- Ask the business stakeholders what is working for them now and what is not. Ask about workarounds, inefficiencies, duplications, pain points, and missing support for their processes. Ask about what they wish were true. Develop requirements that speak to the resolution of these issues, or the realization of what is wished to be true.

**They keep changing their minds.**

1) A Possible Root Cause

   - The business changes.

2) New Truth

   - I can conduct analysis knowing that some things are more stable than others.

3) Possible Response Strategies

   - Develop skill in distinguishing what is likely to remain stable as a core function of the business from what is likely to change as the business changes.

   - Develop skill in techniques that capture changing business rules, such as with the use of decision tables. (This information will help technical staff bring solutions that adapt gracefully to a changing business, such as through the use of business rule engines or storing changes as updatable table values rather than as code.)

- Ensure that a requirements change management process and a document version control process are in place to manage and track changes.

- Set your expectations to expect change.

**I don't have enough access to actual users or subject-matter experts.**

1) A Possible Root Cause

    - People are too busy to meet with me.

2) New Truth

    - People make time for me because of the value I bring.

3) Possible Response Strategies

    - Consider each person you meet with uniquely. How can you speak to his or her situation?

    - Articulate the purpose of every meeting and what will be produced.

    - Come prepared and be sure to capture information so that you are not asking the same questions again.

    - Offer to meet at the other person's convenience, such as off-hours or over lunch. Be accommodating. The priority of a meeting with you will rise as you show value.

# INDIVIDUAL CHALLENGES AND POSSIBLE RESPONSE STRATEGIES

4) Another Possible Root Cause

- I am blocked from access to SMEs because others play this role, yet I am the BA.

5) New Truth

- I have a clear, distinct BA role, and in this role I have adequate access to SMEs.

6) Possible Response Strategies

- Discuss issues with overlapping roles with decision makers as a first step. It could be that they are unaware of these issues.

- Provide references that support that the BA is the key liaison with SMEs when it comes to analysis.

- Detail the difficulties in completing your work *without* access to SMEs, and the limits of what is provided to you by those *with* access to SMEs. Specific examples will illuminate the issues expressed.

- Discuss the "burn factor." When others burn time with SMEs analyzing an area of the business and then provide inadequate documentation to the next analyst, the time is burned. The business is less willing to go through the same discussion again with someone else.

**The stakeholders are not on the same page.**

1) A Possible Root Cause

    - The stakeholders have different, competing interests, and their goals have drifted apart.

2) New Truth

    - The stakeholders remain on the same page throughout the project.

3) Possible Response Strategies

    - The project changes, the world changes, and the vision for the project should be revisited along the way. Bring the stakeholders together on a regular basis to revisit their project vision.

    - Sometimes when people agree, they agree to different parts of what was discussed. Bring the stakeholders together to make sure the former agreement is understood in the same way, and make changes if necessary.

**This takes more time than anyone knows.**

1) A Possible Root Cause

    - People don't know the time it takes to conduct analysis.

2) New Truth

- People understand the time it takes to conduct analysis.

3) Possible Response Strategies

- Consider donating time if it is to showcase what can be done. This is risky, as you don't want to set yourself up to do this on a regular basis. But this is a way to demonstrate value first to provide the rationale for more time the next time. Be sure to share that donated time was part of making the effort a success.

- Document your time as you conduct analysis, and bring this information to the table when estimating time for new similar analysis.

- Partner with those who need to be more aware of the process, and share information about your business analysis work step by step.

- Provide effort estimates with your approach, along with the basis for your estimates, to set expectations up front. Adjust the estimates and communicate the adjustments as you know more.

# WORKING WITH THE TECHNICAL STAFF IS PAINFUL: CHALLENGES AND POSSIBLE RESPONSE STRATEGIES

> BA 1: How did you find time to finish your novel?
>
> BA 2: Easy. I found time by reusing the last project's requirements package and changing the title to my new project. They never read it anyway.

**I find it impossible to communicate with the technical staff.**

1) A Possible Root Cause

    - I don't understand the "techno-babble."

2) New Truth

    - I have a basic understanding of technology.

3) Possible Response Strategies

    - Get some technical training.

- Ask for the technical staff to explain what you don't understand, if it is important that you understand.

- Have a discussion about the handoffs and how you can smooth the transition from analysis to design (or the transition point that applies in your situation). Ask for clarification of technical information in this context.

4) Another Possible Root Cause

- The technical staff dismisses me.

5) New Truth

- The technical staff views me as a professional who brings value to the team.

6) Possible Response Strategies

- Leverage a breakthrough with one technical staff person to demonstrate and/or document how business analysis brings value.

- Build rapport with the technical staff.

- Discuss the intersection of the BAs and the technical staff as a team.

- Articulate the risks in not addressing analysis properly.

- Analyze the team process and outputs. Determine if you can contribute BA knowledge and skill to meet a need or gap.

- Learn more about the technologies being used, and have discussions on the goals and methods of the technical staff.

- Escalate the issue to a supervisor if necessary.

**My work is not used downstream.**

1) A Possible Root Cause

   - They do not believe analysis really matters downstream.

2) New Truth

   - They value and use my analysis downstream.

3) Possible Response Strategies

   - Invite technical staff to your sessions or interviews as observers.

   - Trace the effect of analysis work or the failure to use analysis work to the final business results to make the case for good analysis.

   - Establish with the technical staff up front which analysis deliverables will be produced and their value to the technical staff.

# INDIVIDUAL CHALLENGES AND POSSIBLE RESPONSE STRATEGIES

4) Another Possible Root Cause

- The technical staff does not understand my deliverables or see the value to them.

5) New Truth

- The technical staff understands and values my deliverables.

6) Possible Response Strategies

- Conduct a walkthrough of analysis deliverables.

- Ask if other deliverables would be more valuable, and consider either altering or augmenting the deliverables you prepare.

7) Another Possible Root Cause

- They feel the requirements deliverables are constraining their creativity.

8) New Truth

- They view the analysis deliverables as helpful to their creative work.

9) Possible Response Strategies

- Evaluate whether you are getting into design in your analysis. Consider whether you are *specifying functionality* that could be supported by many solutions, or whether you are *specifying a solution.*

- If your role tips into design, have a conversation with your team about role differentiations. What is analysis and what is design as defined by your team?

- Consider the software methodology in place and whether your analysis is in synch. If not, work it out with the team.

**They have started working on the solution before the analysis is done.**

1) A Possible Root Cause

    - They are pressured to make progress as soon as possible.

2) New Truth

    - Business analysts and developers work well together within the time constraints imposed in the environment.

3) Possible Response Strategies

    - Explore with the technical staff whether other approaches might work better for the environment. For example, a handoff of framing-level business models followed by development of detailed business models and user stories during an Agile iteration might work well, or attacking the work in smaller increments might be the best approach.

- Make the case for why it might take more work to begin development before the analysis is done due to corrections and refactoring. Identify which types of analysis are most vital to finish before development. For instance, discuss why it is important to move from an as-is state to a to-be state with the analysis before introducing technology.

- Document the fallout from development that went forward before analysis completion. Also document the fallout experienced by development because of analysis delays. Have an honest discussion about how to make analysis, design, and development work given the environment.

- Offer to model the as-is process of analysis, design, and development. Together, build a to-be process model.

4) Another Possible Root Cause

- They are working by a software development methodology that conflicts with the business analysis methodology used.

5) New Truth

- The business analysis and software development methodologies are in synch.

6) Possible Response Strategies

- Identify the points of conflict in the two methodologies, and work collaboratively to resolve the conflicts. Evaluate the methodologies used and explore needed changes.

- Consider whether different approaches could be used but still be complementary (e.g., having a plan-driven analysis approach that is followed by a change-driven software development approach, such as Agile).

## The technical staff are bypassing me and going directly to the business/users/customers and now my work is out of date.

1) A Possible Root Cause

- They don't know that bypassing the BA is a problem.

2) New Truth

- The technical staff is aware that all analysis changes must involve the BA.

3) Possible Response Strategies

- Explain the issue to the technical staff, and why it matters (e.g., the analysis gets out of date and the to-be analysis cannot be readily used as the baseline as-is analysis next time).

- Explain why it is important to leverage roles and the expertise of each role. Bypassing the BA means that the analysis might not be updated appropriately or that the solution might not be aligned with the analysis.

4) Another Possible Root Cause

- The technical staff feels it is easier and/or quicker to go to the SMEs directly.

5) New Truth

- The technical staff understands that easier or quicker means compromises to the analysis process and ultimately negative effects on the business.

6) Possible Response Strategies

- Acknowledge that going to the business users or customers is easier and quicker, and then list the tradeoffs. For example, analysis not updated correctly or maintained for reuse, roles are not respected, and the decisions might be out of alignment with the bigger picture.

- Evaluate whether you can speed up your analysis process in regard to requirements changes.

- Discuss the "burn factor": "If you ask the business these questions and you imbed the answers as code, for instance, I am perceived as covering the same territory when I ask the same questions to update the analysis artifacts."

7) Another Possible Root Cause

- They feel that they can understand the requirements better by cutting out the middleperson.

8) New Truth

- The technical staff feels comfortable with the established roles and the level of involvement they have with the business users.

9) Possible Response Strategies

- Consider more collaborative models for working with the technical staff and business persons as a team.

- Invite the technical staff to sessions or interviews as observers, if this fits your environment's culture.

- Demonstrate the expertise and skills you have that they don't, and what is lost if you are not utilized.

## There isn't a professional space for me now that we have gone Agile.

1) A Possible Root Cause

   - The developers conduct the analysis.

2) New Truth

   - I have a role as a BA in this Agile environment.

3) Possible Response Strategies

   - Explore industry standards and practices for working as a BA in an Agile environment.

   - Explore whether analysis can be done outside of Agile projects (i.e., work the backlog to develop business models that can be used during the next iteration or release). Discuss with the team the possibility of developing functional groupings of backlog items, while staying true to the business prioritization of the backlog items. Offer to determine these functional groupings.

   - Become valuable to the team by showcasing the business analysis skills that fill a gap on the team.

4) Another Possible Root Cause

   - I am asked to create user stories and that's it.

5) New Truth

    - I feel comfortable with my role in this Agile environment.

6) Possible Response Strategies

    - Evaluate whether your current role definition is acceptable and feels valuable to your career development. If not, you may need to make some tough career decisions.

    - Explore industry standards and current thinking on user stories. Develop the best user stories possible. Develop other analysis artifacts (e.g., data flow diagrams) in order to produce well-crafted user stories (with high cohesion and low coupling).

    - Explore with your team whether an adjustment in your role is possible (e.g., the incorporation of business models, or serving as a Scrum Master).

    - Explore whether other opportunities in your organization exist, such as enterprise analyst or business architect.

**I feel like I'm leading the business stakeholders to slaughter, as the technical team is not strong.**

1) A Possible Root Cause

    - The technical team has not been challenged to do more than they have done.

2) New Truth

- The technical staff is strong and also understands the importance of business analysis.

3) Possible Response Strategies

- If it seems reasonable, speak with the technology staff about your concerns. Stay focused on how the issues may result in not meeting business or customer needs. Have a frank discussion about which business needs cannot be met and whether anything can be done to meet these needs. If needed, discuss how this inability to meet these requirements should be relayed back to the business or customer and documented as risks.

- Develop strong requirements that hold solutions accountable to them. This applies to both functional and nonfunctional requirements.

- Stress that it is important to identify the "what" before the "how." Once the functional and nonfunctional requirements are clear, a discussion about solution options would be based on *how well* the solutions meet the requirements.

- Escalate your concerns to your supervisor if needed.

# THE MANAGEMENT OF THE PROJECT IS DRIVING ME CRAZY: CHALLENGES AND POSSIBLE RESPONSE STRATEGIES

> PM: Hal, get started on the implementation and configuration of the software we just purchased. Sally, start writing some requirements.

**I am being pressured by the project manager for estimates before I've done any analysis.**

1) A Possible Root Cause

- The project manager needs analysis estimates to complete project estimates up front.

2) New Truth

- I can provide estimates at any point with the qualification of the type of estimates I am providing.

3) Possible Response Strategies

- Discuss the reality of estimates when considering analysis. Analysis is all about discovery, and it is hard to estimate what you have yet to discover.

- Establish a categorization scheme for estimates in collaboration with others (rough estimate, estimate based on past work, estimate based on function point analysis, an estimate range, etc.).

- Research the industry for estimating techniques.

- Get training on estimation.

- Explore with the project manager whether there is latitude on the timeline or precision of the needed estimates.

**I am not given the time I need to do adequate analysis.**

1) A Possible Root Cause

- Those who determined the allotted time don't know how much work this is and how much time I need.

2) New Truth

- I have adequate time to do analysis.

3) Possible Response Strategies

- Provide honest estimates with a basis or the estimates won't be trusted next time.

- Document activities and the time it takes in order to make the case for adequate time on this project or the next one.

- Be transparent and discuss the approach, the amount of time needed, and the purpose of each activity with all who should be consulted.

4) Another Possible Root Cause

- The project is extremely time-pressured.

5) New Truth

- I am able to function well under pressure.

6) Possible Response Strategies

- Evaluate the effect of the project's time pressure on your work. Determine steps you can take to mitigate the effect, such as asking for other projects to be delayed or reassigned.

- Determine the risks and the tradeoffs that result from the time pressure in relation to your work. Then be sure to communicate this, e.g., "I can analyze the process in that amount of time, but it will be at a higher level of granularity than might be needed. The risk is that there may be a 'gotcha' in the details."

- Over time, assess whether you can improve your own processes including your speed, either with team or individual improvements (such as by using templates or standard model fragments based on analysis patterns you see in your organization).

## The project decision makers are making bad decisions that affect my work.

1) A Possible Root Cause

    - The project decision makers lack knowledge and skill in their field of expertise.

2) New Truth

    - I can function well and do the best I can do in my situation despite less than desirable decision making around me.

3) Possible Response Strategies

    - Ask decision makers lots of questions to better understand their decisions and needs. This might shift your understanding or create an opportunity to influence them. Questions are more palatable than assertions to those you wish to challenge.

- Determine how you might mentor those around you strategically (e.g., ask to collaborate on their work so that you better understand the expectations for your work. Look for opportunities to mentor in the process).

- Alter the expectations you have for yourself (i.e., determine what the best work is that you can do given the situation and strive for this goal. Acknowledge to yourself what you cannot control).

4) Another Possible Root Cause

- Project decision makers wish to maintain control and believe they are making correct decisions (even if you feel they are not).

5) New Truth

- I am in synch with the project decisions and feel their decisions are correct.

6) Possible Response Strategies

- Ask questions to better understand.

- Consider whether you can influence the decisions once you understand them better by making the case for how they should be changed. Say, "The risk I see with this decision is this..." in order to articulate why they might make a different decision.

- Look for motivators or drivers of the decision makers to identify how a different decision would be in their best interest, and then relay this information to them.

1) **How can I do analysis when the project is sinking?**

    1) A Possible Root Cause

        - The project does not have the needed resources.

    2) New Truth

        - The project goals and scope are in synch with the resources.

    3) Possible Response Strategies

        - Relay your concerns to the project manager and offer suggestions, based on your analysis, for possible scope adjustments to match the available resources. This might also start the conversation of other options (e.g., securing more resources).

        - If you feel the project is tanking, but you are still actively conducting analysis, make as much progress as possible by advancing your analysis artifacts to be picked up by a future project.

        - Observe the impact of inadequate resources in order to make the case for change and/or to learn.

4) Another Possible Root Cause

- The project manager is not able to do the job.

5) New Truth

- The project is managed by someone with the appropriate skill.

6) Possible Response Strategies

- Speak from the perspective of an analyst interfacing with the project manager's work and provide mentorship. For example, suggest that you review the plan together to clarify your analysis work. Or ask for status updates so that you can evaluate whether your analysis is in synch with the project.

- Consider escalating the issue to the project manager's supervisor if the situation warrants doing so.

- Suggest project management training to the appropriate decision makers in a sensitive way.

- Observe the impact of inadequate project manager skill in order to make the case for changes and/or to learn.

- Offer to help the project manager plan by offering your business analysis skills, such as modeling the full project approach with the project manager serving as the subject-matter expert. Ask questions that fill out the plan, such as "What is your estimate for how long this will take?" "Whom should we ask for an estimate on this task?" "What information or resources are needed to complete this task?"

1) **The purpose of the project is not clear.**

   1) A Possible Root Cause

      - People have different ideas of what we are doing and why.

   2) New Truth

      - People are in synch in terms of the project's objectives.

   3) Possible Response Strategies

      - Bring people together in one room to work through the project vision.

      - Identify points of contention or disagreement. Try to identify whether there are other options that meet all interests, different than what has been expressed so far. Also identify common understandings, which are often best done first to establish common ground.

      - Make use of the question "why?". Not only might this reveal more direct business drivers, it might also reveal goals on which agreement can be gained.

4) Another Possible Root Cause

- We didn't take the time to determine the purpose of the project before we began the work.

5) New Truth

- It is an expectation that a vision must be developed for all projects at the beginning of the project.

6) Possible Response Strategies

- Develop a standard process and template for creating a statement about the vision, charter, definition, purpose, etc. Be sure to consider impacts, alignments, or balances to strike between the project and other projects.

- Draft a vision statement and shop it around to improve it.

- Document the wasted time and effort that is directly related to not having a common vision, such as project activities that are at cross-purposes. Share this information.

# THE BUSINESS ANALYSTS ARE NOT WORKING AS A TEAM: CHALLENGES AND POSSIBLE RESPONSE STRATEGIES

**There is competition among the BAs.**

1) A Possible Root Cause

    - The BAs are in competition to build a strong professional reputation and the team has not "gelled."

2) New Truth

    - The BAs have each other's backs and work as a cohesive, effective unit.

3) Possible Response Strategies

- Call it and have a discussion. "We seem to be in competition. Do you agree? Why are we? What can we do about this?"

- Be the first one to champion the others and support their work.

- Make the case for why one of you doing well means your team is viewed as doing well.

- Take on work as a team, and switch out roles on a rotating basis.

- Engage in team-building exercises on and off site.

## The BAs disagree about the correct approach.

1) A Possible Root Cause

- We are not on the same page in regard to how things should be done.

2) New Truth

- We are in alignment in regard to our analysis methodology.

3) Possible Response Strategies

- Develop a team methodology together, looking to the industry for insights and expert knowledge.

- Have regular reviews regarding how well the methodology is working or not working, and make adjustments.

4) Another Possible Root Cause

- The BA lead has not built faith in the approach.

5) New Truth

- The BA lead has the backing of the team and is successfully providing leadership.

6) Possible Response Strategies

- Establish an approach up front to gain input and agreement among team members. Actively review, evaluate, and update the approach as a team.

- Support the lead by routing ideas through him or her rather than acting on them without keeping the lead in the loop.

- Determine what you can do to lift up each team member you work with. If you are the lead, set the example.

**The roles among the BAs are not differentiated.**

1) A Possible Root Cause

- We haven't taken the time to differentiate roles.

2) New Truth

- The role differentiations among the BAs are clear.

3) Possible Response Strategies

- Talk through fair ways in which the BA roles and responsibilities could be differentiated. For example, distinguish by senior and junior; by business domain; by capability such as facilitator or power user of a BA tool; or by analysis skill such as process or data, etc.

**Other BAs are putting me down outside of our BA team.**

1) A Possible Root Cause

- There are personality differences that get in the way of our ability to work as a team.

2) New Truth

- There are personality differences and we make it work.

3) Possible Response Strategies

- Engage in personality assessment exercises to understand each other, such as the Myers-Briggs Inventory.

- Leverage each BA's strengths and differentiate responsibilities to match.

- Consider what is behind a difficult team personality. Identify that person's needs and attempt to align your efforts to meet those needs.

4) Another Possible Root Cause

- Others sabotage my work to raise themselves up.

5) New Truth

- We respect each other's professional discretion, yet we collaborate and accept constructive criticism. We know that a team in turmoil makes us all look bad.

6) Possible Response Strategies

- Develop ground rules such as bringing concerns back to the team rather than raising concerns in front of SMEs.

- Offer two items of praise for every criticism.

- Ask for feedback and resist being defensive.

- Try to determine what is behind the behavior of the saboteur. If possible, determine what you can do to address the deeper issue. Does he or she feel threatened? Has this person been the only BA and now there is more than one to adjust to?

**There are too many cooks in the kitchen.**

1) A Possible Root Cause

   - Roles and responsibilities are not clear and more than one person steps forward as leader.

2) New Truth

   - Roles, responsibilities, and leadership are clear.

3) Possible Response Strategies

   - Discuss the fallout of having more than one person trying to lead an effort.

   - Have a discussion about roles and responsibilities and clarify who is in charge.

   - Establish a protocol for providing feedback to the agreed-upon leader.

   - Consider whether having different BAs serve as lead on different efforts will work for the team.

# I AM STRUGGLING: CHALLENGES AND POSSIBLE RESPONSE STRATEGIES

> *Requirements:*
> *1. The system must, it must self-combust.*
> *2. The system shall, it shall wreck morale.*

**I am now a BA, but I don't have any training or experience as a BA.**

1) A Possible Root Cause

    - Standards did not exist for business analysis work and I am the result.

2) New Truth

    - I grew into my job, even though I was not prepared when I began.

3) Possible Response Strategies

    - Survey the industry to learn as much as possible about business analysis.

- Connect with the International Institute of Business Analysis (IIBA).

- Set a goal to learn one new technique every week, month, or quarter.

- Find a BA mentor.

4) Another Possible Root Cause

- I signed up for more than I am capable of.

5) New Truth

- I meet the challenges of my BA position.

6) Possible Response Strategies

- Consider whether you need to be more honest about your capabilities. Scale back the expectations others have of you and that you have of yourself.

- Break down a large assignment into a step-by-step approach. Evaluate each step and decide if you need help.

- Identify specifically what you need to have in place to meet a stretch goal (e.g., more training, mentorship, assistance, extra time, etc.).

- Discuss your specific needs for meeting a stretch goal with your supervisor.

- Own up to the situation when you need to get help. It is a strength to ask for help when you need it.

**The work is so varied, I don't know which approach or tools to use.**

1) A Possible Root Cause

    - We don't have any standards in my organization.

2) New Truth

    - We have useful standards in my organization.

3) Possible Response Strategies

    - Make the case for why the lack of standards is a problem, and cite lessons learned and/or show evidence to support this case.

    - Volunteer to research and evaluate industry standards, to develop internal standard recommendations, and/ or to facilitate discussions around standards. Suggest standardization of the business analysis process, templates, tools, approaches, etc. to the appropriate decision maker. Be careful not to overshoot so the BAs aren't able to exercise their professional discretion and customize an approach to fit an analysis effort.

4) Another Possible Root Cause

- I am lacking skill in certain tools or techniques.

5) New Truth

- I am skilled in the techniques and tools that I need to do my job.

6) Possible Response Strategies

- Ask for training.

- Observe other BAs and review their work. Ask that other BAs observe you, review your work, and provide feedback.

- Practice on your own time and analyze areas of your own world. This will benefit your career, not just your current job.

**I am just burned out—this is exhausting.**

1) A Possible Root Cause

- I work too many hours.

2) New Truth

   - I work a reasonable number of hours.

3) Possible Response Strategies

   - Document your activities and hours, even if not asked to do so. Do this in order to communicate with your supervisor and estimate similar efforts.

   - Develop a recommendation for how to resolve the issue in a way that meets your needs, your supervisor's needs, and the organization's needs before talking with your supervisor.

   - Consider whether other options would meet your needs, such as more compensation for overtime hours or relieving you of a project, and make these recommendations to your supervisor.

   - If you are a consultant, build in controls over the work you will take on into your up-front interview and negotiation process rather than agreeing to whatever it takes, if your goal is to constrain hours.

4) Another Possible Root Cause

   - My work is not used.

5) New Truth

    - My work is valued and used.

6) Possible Response Strategies

    - Determine why your work is not used, so that you can make changes. Should your deliverables change? Do others need to understand the deliverables? Have others not bought into the need to even consider your work?

    - Walk through the deliverables with those who will use them, and be open to making adjustments.

    - Develop your knowledge and skill.

    - Consider whether role conflicts are under the surface and address them.

7) Another Possible Root Cause

    - I encounter a lot of conflict with people.

8) New Truth

    - I work well with other people.

9) Possible Response Strategies

- Ask a few trusted co-workers why they think you encounter so much conflict. Consider the feedback carefully and determine whether you can make changes.

- Determine a new approach when working with others and test it out. Then determine another approach and test it out, and so on.

- Focus on the needs of the other person and ask a lot of questions.

10) Another Possible Root Cause

- I am bored.

11) New Truth

- I feel fulfilled by my work.

12) Possible Response Strategies

- Ask for more work.

- Ask for new types of work.

- Ask for more training.

- Evaluate how you could do more advanced work with current assignments and discuss this with your supervisor.

- Mentor others.

- Have a long-term goal and plan for your career.

- Search for new opportunities.

# ABOUT THE AUTHOR

**LeAnn Simonson** is a Certified Business Analysis Professional (CBAP), practicing business analyst, consultant, author, and teacher. LeAnn has been a software developer and a social worker. She holds master's degrees in software engineering and in social work, both from the University of Minnesota, where she is an adjunct professor. LeAnn and her family reside in Minnesota, where she is occasionally accused of being too "Minnesota nice."

Printed in Great Britain
by Amazon.co.uk, Ltd.,
Marston Gate.